Regression Analysis for Categorical Moderators

Methodology in the Social Sciences
David A. Kenny, *Series Editor*

Regression Analysis for Categorical Moderators

Herman Aguinis

SERIES EDITOR'S NOTE by David A. Kenny

THE GUILFORD PRESS
New York   London

© 2004 The Guilford Press
A Division of Guilford Publications, Inc.
72 Spring Street, New York, NY 10012
www.guilford.com

Printed in the United States of America

This book is printed on acid-free paper.

Last digit is print number: 9   8   7   6   5   4   3   2   1

Library of Congress Cataloging-in-Publication Data

Aguinis, Herman, 1966–
  Regression analysis for categorical moderators / by
  Herman Aguinis.
     p. cm. — (Methodology in the social sciences)
    Includes bibliographical references and index.
    1-57230-969-5
    1. Regression analysis. 2. Social sciences—Statistical
methods—Data processing. 3. Regression analysis—
Computer programs. I. Title. II. Series.
  QA278.2.A29 2004
  519.5′36—dc22

                           2003024141

To my wife, Heidi

The most important moderator of the relationship between my life and my happiness

About the Author

Herman Aguinis, PhD, (*www.cudenver.edu/~haguinis*) is Associate Professor and Director of the Management Programs at the University of Colorado at Denver. He has held visiting appointments at China Agricultural University (Beijing, China), City University of Hong Kong (China), University of Science Malaysia (Penang, Malaysia), and University of Santiago de Compostela (Spain). He has published over 40 articles in refereed journals and delivered over 100 presentations in the United States, Asia, Europe, and Latin America on the topics of research methods and statistics, personnel selection, and social power and influence in organizations. He currently serves as Associate Editor for *Organizational Research Methods* and on the editorial board of several journals, including *Journal of Applied Psychology* and *Journal of International Business Studies*. He has been elected Chair of the Research Methods Division of the Academy of Management for 2003–2004.

Series Editor's Note

This book deals with a very common problem in both basic and applied research. In basic research, researchers often have an independent variable and they seek to know if the effect of the independent variable changes as a function of some continuous variable (e.g., age or a baseline score). That is, they are interested in understanding the limits of the effect of the independent variable on the dependent variable. In applied research, there is often interest in the ability of a variable to predict another variable. However, there is the concern that the prediction function may vary by gender or race. In both cases there are two variables, one categorical and the other continuous, that explain an outcome variable. The concern is that the two interact to explain the outcome. Alternatively and more commonly, it is said that one variable, usually the categorical variable, *moderates* the relationship between the other two variables.

Interest in this type of question is not just academic. Employers often use tests to make decisions to hire workers. They have prior evidence that a test predicts worker productivity. The following question is very important: Does the test predict the productivity of men and women equally well or of Whites and minorities equally well? The answer has important implications for the employer, the test makers, potential and current employees, lawyers and judges involved in discrimination lawsuits, and the general public. The answer to the moderation question can affect people's lives and can cost millions of dollars.

In this book, Herman Aguinis shows that there is a straightforward approach to the moderation question, which is to perform moderated multiple regression analysis, MMR. He presents a variety of contexts in which one would apply MMR and explains the relevant equations, graphs, and computer software.

Although questions involving moderator variables are common and in some ways easy to analyze, there are many subtle problems that can make the estimation and interpretation of results much more complicated. This book details those problems. First and foremost is the problem of power. To have a reasonable chance of yielding a statistically significant result, tests of moderation often require prohibitively large sample sizes. It is quite common for sample sizes on the order of 800 cases to be needed. This problem of low power often went unrecognized in the past, as shown by the reviews of previous literature mentioned in this book. Students planning dissertations in which they wish to test moderator hypotheses would be well advised to study carefully Aguinis's analysis of the low-power problem. No one wants to spend a year of his or her life doing a study that has only a 20% chance of working.

A host of other problems and issues are explored in the book. Here are some: Prediction errors have different variances group to group, mathematical functions for the relationship between the predictor and the outcome are often nonlinear, there are alternatives to dummy coding of the categorical variable that should be considered, and there is risk of measurement error in the predictor variable—to name just a few. For each of these, Aguinis clearly details the problem, illustrates it with a data set, and then gives explicit strategies for the solution to the problem. Often a computer program would be of great assistance in finding these solutions; Aguinis has written several such programs and has made them generally available (on the Web site *www.cudenver.edu/~haguinis/mmr*). I strongly urge readers to download the data sets that Aguinis discusses and redo his analyses, as well as to try out the methods detailed by him on their own data.

Although some of the topics that Aguinis discusses are complex, he explains these difficult problems using plain language, instead of resorting to the esoteric Greek formulas that we methodologists love to use— yet he still manages to detail that complexity. He offers a rich array of examples from various disciplines in the social and behavioral sciences to illustrate nuances in the solutions to MMR problems, surgically dissecting them, and providing researchers with clear and explicit guidance to the detection of and the answers to the moderation question.

DAVID A. KENNY

Preface and Acknowledgments

Consider the following questions:

- Does the prediction of grade-point average (GPA) based on SAT scores vary across types of college applicants (e.g., private vs. public high school graduates)?
- Is a lunch-session psychotherapy intervention for teenagers with anorexic symptoms more successful for girls as compared to boys?
- Is a new performance management system more successful at motivating managers as compared to entry-level employees?
- Does the relationship between leader behavior and team performance vary across cultures?
- Does the process of setting goals (e.g., specific vs. "do my best") lead to different levels of goal attainment depending on the level of goal difficulty?
- Do birth-order effects vary across ethnic groups?
- Does the effect of promotion on sales depend on the market segment under consideration?

The answer to each of these questions implies a contingent or moderated relationship where the causal effect or association between two variables changes across discrete categories. In other words, a moderated relationship gives us information about when, or under which conditions, the relationship between two variables is likely to be stronger or weaker. Moderated relationships provide more detail than main effects regarding the phenomena under consideration and, therefore, give us a more fine-tuned picture of reality. This is why the extent to which researchers understand moderated relationships is at the heart of the scientific enterprise. Moreover, an understanding of moderated relationships is often regarded as an indicator of a field's scientific advancement and maturity. From a practical standpoint, an understanding of moderators allows researchers to plan and implement interventions that are ethical, fair, and legally appropriate for the various groups under consideration.

Moderated multiple regression (MMR) is a statistical tool of choice to conduct moderator analyses of grouping variables in the social sciences. However, an MMR analysis is much like a minefield: There are numerous unknown factors that threaten the accuracy of MMR results. Furthermore, these often unknown threats are likely to render MMR-based conclusions invalid. Thus, MMR users must be "mindful in the minefield."

The purpose of this volume is to provide practical guidance to social and behavioral scientists interested in using MMR to assess the classical question of whether the relationship between two quantitative variables is moderated by group membership. As will become evident, testing a hypothesis regarding moderating effects of *categorical* variables presents unique challenges as compared to testing for moderating effects of *quantitative* variables. Some of these unique challenges include how to choose a coding system, how to interpret the resulting regression coefficients, and the effects of unequal sample sizes and error variances across moderator-based subgroups. Thus, although some of the issues discussed in the book also apply to tests of quantitative moderators (e.g., the central role of theory, low statistical power), the book focuses on categorical moderators.

The book includes a discussion and fully worked-out examples of how to conduct and interpret MMR analyses as well as descriptions of computer programs that allow investigators to check whether their MMR tests for moderation can be trusted. Because of its practical nature, the book and its associated computer programs and data sets can also be used as supplementary materials in advanced undergraduate statistics and methods courses as well as graduate courses addressing the general linear model. This volume includes only the essential mathematical material and equations in text because the goal is not to provide proofs or the mathematical underpinnings of moderator analysis. The assumption is, however, that readers have some knowledge of descriptive and basic inferential statistics and multiple regression. The organization of the book is as follows:

Chapter 1 defines moderated relationships and provides examples from several social sciences (e.g., psychology, management, education, political sciences, international business) to illustrate the importance of moderated relationships for science and practice. This chapter also distinguishes moderated from mediated relationships and describes the importance of having a sound rationale before conducting a moderator analysis.

Chapter 2 presents a description of the MMR procedure and its basic statistical assumptions. This chapter also summarizes evidence demonstrating the appropriateness of using MMR to assess moderating effects. The conclusion regarding the appropriateness of using MMR is supported by numerous independent studies and by the fact that the technique has the endorsement of several professional and scientific associations. Finally, it describes the pervasive use of MMR in the social sciences to determine whether a moderator variable exists.

Chapter 3 provides a detailed description of how to conduct an MMR analysis using computer packages and how to interpret the resulting output. Although this chapter focuses on SPSS, it provides a description of all the logical steps involved so that they can be implemented with other widely available software packages. This chapter includes a description of a specific research scenario and the data set available on the World Wide Web (hereafter referred to as the Web) to be used in the MMR analysis. (The Web address for this data set and the other MMR programs and resources I've developed is *www.cudenver.edu/~haguinis/mmr.*) The material in this chapter can best be understood by a hands-on approach, so, ideally, readers should follow the steps outlined in the chapter in front of a computer.

Chapter 4 describes the homogeneity of error variance assumption and differentiates it from the homoscedasticity assumption. The homogeneity of error variance assumption is a critical component of the MMR model, and this chapter shows the effects of violating the assumption on substantive research conclusions. Also, this chapter describes a computer program available on the Web (including fully worked-out examples) that allows researchers to (1) check whether the assumption has been violated and (2) conduct a test for moderation using analyses other than MMR that allow for heterogeneity of error variance.

Chapter 5 discusses statistical power and frames the issue within the broader topic of statistical inference. First it describes briefly the controversy over null hypothesis significance testing (NHST). Then it provides a nontechnical yet detailed description of the numerous factors that affect the power of MMR. These factors include total sample size, preset Type I error rate, moderating effect magnitude, predictor variable variance reduction, error variance heterogeneity, measurement error, and scale coarseness, among others. The chapter makes clear that the conditions that are adverse to adequate statistical power are unfortunately very common in most social science research settings where MMR is used. Finally, this chapter reports results of a 30-year review describing the small magnitude of typically observed effect sizes and the relationship between effect sizes and power in published research.

Chapter 6 offers several suggestions on how to minimize the adverse effects on power of each of the methodological and statistical artifacts discussed in Chapter 5. The proposed solutions address issues related to design, measurement, and data analysis. The chapter discusses a range of solutions broad enough that all MMR analyses can benefit from the implementation of at least some of them. Some of the proposed solutions are easier and more cost-effective to implement than others. All MMR users who wish to minimize the impact of the factors likely to lead to the incorrect conclusion that there is no moderating effect should seriously consider using the suggested strategies that help increase power.

Chapter 7 discusses the importance of determining statistical power for a study, then describes computerized tools that allow researchers to know the power of an MMR test. Specifically, the chapter provides a step-by-step analysis of power computation using specific examples of how to use three available programs, two of which are based on empirical (i.e., simulation) results and the third on analytic work. Finally, the chapter makes the case that it is no longer justified to conclude that the null hypothesis of no moderating effect is correct unless a power assessment has resulted in a value sufficiently high to detect a hypothesized moderating effect.

Chapter 8 addresses MMR models that are more complex in nature than the more typical MMR model. Specifically, this chapter describes how to (1) set up four different coding schemes to analyze data sets including moderator variables with more than two levels, (2) analyze and interpret moderating effects vis-à-vis nonlinear effects (e.g., quadratic, cubic), and (3) test and interpret MMR models including three-way and higher-order interaction effects.

Chapter 9 includes a discussion of further issues in the interpretation of moderating effects. Specifically, this chapter includes a discussion of how to compute each of seven indicators of the "practical significance" of a moderating effect. Also, this chapter discusses additional issues related to interpretation, including the signed coefficient rule, and the importance and consequences of not specifying clearly which variable is the criterion and which variable is the moderator in the MMR model.

The final chapter provides an integrated summary and overall set of conclusions for the entire book. It also summarizes guidelines and recommendations for the use and interpretation of MMR to estimate moderating effects of categorical variables. Among other topics, the chapter emphasizes the importance of theory in the process of searching for moderators, the benefits of computing statistical power before a study is conducted, the need to take into account the coding scheme and data transformations implemented (such as centering in interpreting results), and the choices available regarding indicators of practical significance.

The hope is that this volume will be beneficial to social scientists and students who seek to understand research questions that are likely to include the phrase "it depends" as part of the answer. Being "mindful in the minefield" will allow researchers to draw accurate conclusions from their MMR analyses. This mindfulness ideally will promote a better understanding of the effects of moderator variables and the development of more sophisticated and fine-tuned models in the social sciences. The hope is also that this book will be helpful to practitioners who seek to understand whether a particular intervention is likely to yield dissimilar outcomes for members of various groups. This is particularly important in such areas as testing in preemployment and educational settings, where conclusions based on MMR have important ethical and legal implications.

I would like to thank several individuals who have been instrumental in the completion of this book. First, I thank Dr. David Kenny for suggesting that I write this book and for his insightful feedback on each of the chapters. Dave: I thank you for your incisive and detailed comments that allowed me to make numerous improvements throughout. Of course, any errors are mine alone. Second, I thank three colleagues with whom I have conducted research on moderator analysis over the past decade: Dr. Charles A. Pierce, Dr. Eugene F. Stone-Romero, and Dr. Robert J. Boik. Chuck, Gene, and Robert: I thank you for the opportunity to have worked together; our collaboration has been critical in shaping the ideas presented in this book. Third, some of the chapters in this book are based on workshops I delivered at the City University of Hong Kong and at the Academy of Management meetings in Washington (2001), Denver (2002), and Seattle (2003). I thank the participants in each of these workshops for the ideas, questions, and suggestions of theirs that helped me improve the pedagogical aspects of this book. Finally, I thank my parents, Marcos and Marita (deceased) Aguinis. Pa and Ma: You taught me to be persistent, to never give up, and to follow my dream; I used every bit of your advice in writing this book.

Contents

8 • Complex MMR Models   117

9 • Further Issues in the Interpretation of Moderating Effects   138

10 • Summary and Conclusions   155

Web address for the author's MMR computer programs
and resources: *www.cudenver.edu/~haguinis/mmr*

1

What Is a Moderator Variable and Why Should We Care?

If we want to know how well we are doing in the biological, psychological, and social sciences, an index that will serve us well is how far we have advanced in our understanding of the moderator variables of our field.
—HALL AND ROSENTHAL (1991, p. 447)

The promising results obtained to date indicate that further work should be conducted in the controversial research area of moderator variables.
—CHEEK (1989, p. 281)

The advancement and theoretical sophistication of the social sciences have motivated researchers to go beyond first-order effects and understand <u>moderated, also labeled interactive, relationships.</u> More specifically, researchers are interested in testing <u>whether the relationship between two variables changes depending on the value of a discrete grouping variable.</u> For instance, social scientists have recently investigated such diverse questions as the following:

- Does the relationship between conflict with parents and depressive symptoms change based on adolescents' nationality (i.e., United States vs. China) (Greenberger, Chen, Tally, & Dong, 2000)?
- Does a firm's risk aversion affect its attractiveness to potential employees differently depending on firm ownership type (i.e., state-owned, international joint venture, or wholly owned foreign enterprise) (Turban, Lau, Ngo, Chow, & Si, 2001)?

1

- Does the relationship between perfectionism and bulimic symptoms in female college students change depending on perceived weight status (i.e., overweight vs. not overweight) (Vohs, Bardone, Joiner, Abramson, & Heatherton, 1999)?
- Does juggling of work and family roles affect self-reports of negative affect and calmness differently depending on the setting of the activities (i.e., work vs. home) (Williams & Alliger, 1994)?
- Does a preemployment test exhibit predictive bias such that the relationship between test scores and measures of job performance depends on ethnicity (e.g., minority or majority) and gender (i.e., male or female) (Rotundo & Sackett, 1999; Saad & Sackett, 2002; Society for Industrial and Organizational Psychology [SIOP], 1987; Te Nijenhuis & Van der Flier, 1999)?
- Does psychosis affect depression differently for various age groups (e.g., 18–39 vs. 40–59, or 60–79) (Jorm et al., 2000)?
- Does the relationship between personal goals and job performance change based on type of goal (easy vs. hard) (Tubbs, 1993)?
- Does the relationship between the strategy an importer chooses to use and the renewal of an importing contract depend on the country of origin of the importer (Peru vs. United States) (Marshall & Boush, 2001)?
- Does the relationship between proactive job search and long-term mental health among unemployed individuals change based on reemployment status (reemployed vs. unemployed) (Wanberg, 1997)?
- Do the effects of advertising content on brand attitude vary across levels of brand loyalty (e.g., high vs. low) (LeClerc & Little, 1997)?

Each of the preceding questions shares the same interest in whether the (presumably causal) relationship between two quantitative variables X and Y changes based on the value of a third discrete grouping variable Z. This third variable Z is labeled a moderator of the relationship between variables X and Y when the nature of this relationship is contingent upon Z (Stone, 1988; Stone-Romero & Liakhovitski, 2002; Zedeck, 1971). For example, a moderating-effect hypothesis could be that the effect of setting personal goals (i.e., X) on job performance (i.e., Y) depends on the level of goal difficulty (i.e., Z, easy vs. hard) such that the relationship is stronger for hard as compared to easy goals. Note that one can also describe the moderating effect as an interaction between X and Z. Furthermore, because interactive relationships are symmetrical, one could refer to the moderating effect of Z on the X–Y

relationship, or to the moderating effect of X on the $Z-Y$ relationship. Which variable is chosen as the moderator depends on the substantive research question.

The grouping moderator variable can be experimentally manipulated (e.g., goal difficulty) or naturally occurring (e.g., gender). For example, it may be the case that the "easy goal" group exhibits a strong and positive relationship between personal goals and job performance whereas the "hard goal" group shows a positive but weaker relationship. Alternatively, it could be that the hard goal group shows no goal–performance relationship whatsoever, or even a strong and negative relationship. Thus, in general, the grouping variable Z is a moderator when the $X-Y$ relationship is not the same across the groups under consideration (e.g., easy vs. hard goal, men vs. women, United States vs. China nationals, and so forth). Note that the nature of the relationship between X and Y is such that X may cause Y (in the case of experimental designs), or X and Y may covary (in the case of nonexperimental designs). In general, however, it is typically assumed that X is a causal antecedent to Y.

Although the interest in moderated relationships has increased dramatically over the past 20 years, social scientists have noted the existence of moderator variables for almost half a century (Abelson, 1952; Edwards, 1954; Frederiksen & Melville, 1954; Gaylord & Carroll, 1948; Ghiselli, 1956; Saunders, 1955, 1956). The label "moderator variable" seems to have been used first by Saunders in 1955 but, according to Zedeck (1971), the concept had been discussed previously. For example, Court (1930) used "joint causation," Gaylord and Carroll (1948) used "population control variable," and Frederiksen and Melville (1954) used "subgrouping variable." Even after Saunders (1956) published his paper "Moderator Variables in Prediction," there were authors who continued to use terms other than that to describe moderating effects. Ghiselli (1956, 1960a, 1960b) used "predictability variable," Toops (1959) used "referent variable," Grooms and Endler (1960) used "modifier variable," Johnson (1960) used "homologizer variable," and Velicer (1972) used "heterogeneous regression." At present, although other terms are used occasionally (Sharma, Durand, & Gur-Arie, 1981), moderator appears to have become the accepted term in most social science disciplines.

WHY SHOULD WE STUDY MODERATOR VARIABLES?

Moderator variables are playing an increasingly important role in social science research and practice. On the one hand, researchers may look for moderators to ascertain whether a causal law is general. In situations where researchers seek to find generalizability, the ideal outcome is the finding that there are no moderators (Hunter, Schmidt, & Rauschenberger, 1984). On the other hand, researchers may look for moderators in an attempt to improve the fit of their models, given that main effects alone may not provide sufficient accuracy in prediction. In these situations, the ideal outcome is the finding that there are strong moderated relationships. Regardless of the outcome, the study of moderator variables has implications for both theory and practice because it provides information on the boundary conditions for the relationships of interest.

Consider the situation in which a researcher finds a positive and very strong effect between preemployment test scores and job performance in a sample of men and a weak and almost nonexistent relationship in a sample of women (cf. Stricker, Rock, & Burton, 1993; Young, 1994). By not identifying the moderating effect of gender and considering the overall test scores–performance relationship only, the researcher would conclude that there is a moderate relationship between test scores and performance (i.e., the average test scores–performance relationship across the two groups). Similarly, the British writer and politician Benjamin Disraeli (1804–1881) noted the following (Huff, 1954): "A man eats a loaf of bread, and another man eats nothing; statistics is the science that tells us that each of these men ate half a loaf of bread." In the aforementioned preemployment testing situation, a similarly incorrect conclusion is that there is a moderate relationship between test scores and performance, whereas in actuality the relationship is positive and very strong for men and virtually nonexistent for women.

From a theory point of view, the erroneous conclusion about the absence of a moderating effect of gender precludes the researcher from understanding the sources of the differential relationship across groups (Baker & Yardley, 2002). For example, are testing procedures implemented in a discriminatory way? Is one gender-based sample more homogeneous than the other? Are there meaningful psychological differences in how men and women interpret the test in question? Failure to understand these differential relationships prevents researchers from learning about boundary conditions for the causal effects in question, and therefore is likely to delay the advancement of theory.

From a practice point of view, using this particular test for personnel selection purposes is likely to lead to errors in prediction (underprediction of future performance for men and overprediction of future performance for women). In turn, these errors in prediction are likely to have detrimental effects for organizational productivity as well as employee satisfaction and well-being, not to mention potential litigation. In short, not understanding the moderating effect has important implications for both theory and practice.

DISTINCTION BETWEEN MODERATOR AND MEDIATOR VARIABLES

It is important to distinguish a moderating from a mediating effect. These effects are distinct and are not necessarily mutually exclusive (e.g., Sheeran & Abraham, 2003). A variable is a mediator of an X to Y relationship when it accounts for the causal relation between X and Y (Baron & Kenny, 1986). Mediators are also called "intervening" or "process" variables because they explain the relationship between two variables. In other words, mediators give us information on *why* or *by what mechanism X* causes *Y* (Frone, 1999). In contrast, a moderator variable explains changes in the nature of the X to Y effect. That is, a moderator explains *when* or *under what conditions X* causes *Y* (Frone, 1999). As noted by Baron and Kenny (1986), "whereas moderator variables specify when certain effects will hold, mediators speak to how or why such effects occur" (p. 1176). Figure 1.1 shows these relationships graphically. The top panel shows a moderated relationship and the bottom panel shows a mediated relationship.

As an example, consider the literature on work stress and alcohol use (Frone, 1999; see Sheeran and Abraham, 2003, for an example in the social psychological literature). Researchers are interested in investigating

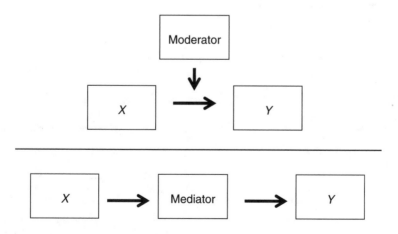

FIGURE 1.1. Representation of a variable serving as a moderator (top) and mediator (bottom).

the effect of work stressors (i.e., aversive work conditions) on employee alcohol use. Some work stressors commonly investigated include dangerous work conditions, noxious physical work environments (e.g., noise and dirt), interpersonal conflict with supervisors or coworkers, and unfair treatment regarding pay (Frone, 1999). Researchers have tested moderated relationships in an effort to understand which are the groups of employees for whom the effect of work stressors on alcohol use is stronger. For instance, work stressors were found to have an effect on alcohol use only for those employees whose work role is psychologically important to their self-definition. On the other hand, for employees whose work is not psychologically important, there was no effect of work stressors on alcohol use (Frone, Russell, & Cooper, 1997). In addition to having an interest in moderated relationships, researchers have also investigated hypothesized mediated relationships in an effort to understand the underlying mechanisms responsible for the effect of work stressors on alcohol use. For example, anxiety was found to mediate the relationship between the work stressor "poor relationship with supervisors and coworkers" and average weekly alcohol consumption (Vasse, Nijhuis, & Kok, 1998). The top part of Figure 1.2 graphically displays the moderating effect of work role centrality (i.e., high vs. low), and the bottom displays the mediating effect of anxiety. This book focuses on moderating effects. Readers interested in mediation should consult Baron and Kenny (1986), James and Brett (1984), Judd, Kenny, and McClelland (2001), and Shrout and Bolger (2002).

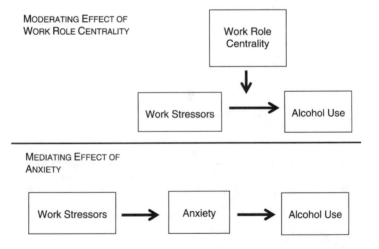

FIGURE 1.2. Representation of work role centrality as a moderator variable (top) and anxiety as a mediator variable (bottom).

IMPORTANCE OF A PRIORI RATIONALE IN INVESTIGATING MODERATING EFFECTS

In many cases, researchers have specific theory-based predictions to guide hypotheses about potential moderator variables (Chaplin, 1997). For example, Greenberger et al. (2000) hypothesized that nationality (i.e., United States vs. China) would moderate the link between the quality of family relationships and depressed mood such that the relationship would be stronger for a Chinese sample as compared to a U.S. sample. This prediction was based on the importance of harmonious relationships, especially within one's family, in Chinese society (Aguinis & Roth, 2003). Because of the high importance placed on harmonious family relationships, there was a strong theory-based rationale to predict that the link between quality of family relationships and depressed mood would be stronger for Chinese as compared to U.S. participants.

In other cases, researchers routinely test for the moderating effects of such variables as gender and ethnicity because tests of that kind are recommended as best practices in specific research domains (Bartlett, Bobko, Mosier, & Hannan, 1978; Bartlett & O'Leary, 1969). For example, over 30 years ago, Einhorn and Bass (1971) issued the recommendation that "it is *always* necessary to investigate the possibility of different regression functions for different groups, since otherwise one might be guilty of discrimination in the use of tests" (p. 263). More recently, the *Principles for the Validation and Use of Personnel Selection Procedures* (SIOP, 1987) recommended such investigations to gather evidence regarding a test's fairness for various groups (Jones, 1973). Thus, although there may not be a specific theory-related rationale for the moderator test, professional practice dictates that this analysis be conducted.

In yet a third type of situation, however, researchers choose to examine the potential moderating effects of grouping variables such as gender and ethnicity without a strong rationale for such investigations. Given that the data have been collected and that conducting a moderator analysis is just a few mouse clicks away, many researchers are tempted to conduct such exploratory analysis in the absence of any type of hypothesis or justification. It may even be the case that a researcher finds an unexpected moderating effect. Although it is permissible to conduct such exploratory examinations, results should be interpreted with great caution and should be replicated using independent samples. As described in more detail in Chapter 5, unless certain conditions exist, it is likely that moderating effects will go undetected in a sample even if they do exist in the population. Moderator variables are difficult to detect even when the moderator test is the focal issue in a research study and a researcher has designed the study specifically with the moderator test in mind. If a moderator test is conducted without a clear a priori justification, based on theory or recommended professional practice, it is likely that the research design will not be conducive to assessing the moderating effect accurately. In short, an a priori rationale should guide the investigation of moderator variables. Otherwise, the conclusions regarding moderating effects may be erroneous.

CONCLUSIONS

■ This chapter defined the concept of a moderator variable, illustrated the pervasiveness and importance of moderated relationships in several social sciences, distinguished moderator from mediator variables, and discussed the importance of having good a priori justification for studying moderated relationships.

■ If the relationship between two variables is not the same across the various groups under consideration, the grouping variable is a moderator (i.e., there is an interaction effect). Social scientists must have a good understanding of the moderator variables in their fields. Otherwise it is likely that resources will not be used wisely and interventions will lead to unintended, and even opposite, outcomes. For instance, does the implementation of a performance appraisal system based on individual goals lead to the same level of performance improvement across cultures? Not having a good understanding of this issue may lead corporate headquarters to implement the same system in all subsidiaries worldwide. Although positive outcomes may be the result in an individualistic culture like the United States, negative outcomes—including decreased productivity, job dissatisfaction, and a deterioration of the level of trust with headquarters—are possible in a collectivistic culture like China (Aguinis & Roth, 2003).

■ Although the focus of this book is on moderator variables, researchers must be aware of the distinction between moderator and mediator variables. Moderator variables provide information regarding the conditions under which an effect or relationship is likely to be stronger. Mediator variables provide information regarding the mechanisms likely to be responsible for the effect or relationship in question. Moderators and mediators are not mutually exclusive, and a hypothesized model may include both (e.g., Frone, 1999).

■ This chapter has also emphasized the importance of having good justification before launching into the analysis of moderator variables. Because analysis of moderators is like a minefield, filled with difficulties that are typically unknown to the researcher conducting the analysis, going on a "fishing expedition" in search of moderators is likely to lead to conclusions that are not replicable in subsequent studies.

■ Chapter 2 describes how to conduct a moderator analysis using moderated multiple regression (MMR) in a nontechnical way. It also describes recently published research in several social sciences that shows how pervasive MMR is for testing whether a grouping variable moderates the relationship between two quantitative variables.

2

Moderated
Multiple Regression

It now seems clear that moderated regression analysis is the appropriate inferential procedure when the underlying theory postulates differences in the form of a relationship between two variables as a function of some moderator variable.
—CHAMPOUX AND PETERS (1987, p. 252)

The previous chapter defined the concept of a moderator variable, provided illustrations of moderators in the social sciences, showed the difference between moderators and mediators, and discussed the importance of having good a priori justification for studying moderated relationships. This chapter describes how to test for the presence of a moderator variable using moderated multiple regression (MMR). The chapter includes a description of the MMR procedure, discusses the appropriateness of using MMR to assess moderating effects, and describes the pervasive use of MMR to determine whether a moderator variable exists. Chapter 3 then provides a step-by-step demonstration of how to implement MMR using computer programs such as SPSS.

WHAT IS MMR?

MMR is an inferential procedure which consists of comparing two different least-squares regression equations (Aiken & West, 1991; Cohen & Cohen, 1983; Jaccard, Turrisi, & Wan, 1990). An inferential procedure would not be needed if a researcher had access to the entire population of true scores (i.e., free of measurement errors). Obviously, this is a rare event in the social sciences. Thus, researchers collect sample data and then make inferences about populations. MMR allows researchers to make the inference of whether a moderating effect is present in the population based on sample data.

Given that a researcher collects data regarding a quantitative criterion or dependent variable Y, a predictor X, and a second binary predictor Z hypothesized to be a moderator, Equation 2.1 shows the ordinary least-squares (OLS) regression equation that tests a model predicting Y from the first-order effects of X and Z:

$$Y = a + b_1 X + b_2 Z + e \qquad \mathit{OLS\text{-}eq}. \quad (2.1)$$

where a is the least-squares estimate of the intercept, b_1 is the least-squares estimate of the population regression coefficient for X, b_2 is the least-squares estimate of the population regression coefficient for Z, and e is a residual term. Note that the criterion Y is a quantitative variable; other procedures such as logistic regression can be used in situations where Y is categorical (Ganzach, Saporta, & Weber, 2000; Jaccard, 2001). Also, Equation 2.1 assumes the prototypical situation where the moderator is binary (i.e., has two categories). Chapter 8 addresses more complex MMR models, including those with moderators with three categories. Equation 2.1 is represented graphically in Figure 2.1.

In Equation 2.1, the regression coefficient b_1 is interpreted as the number of units that Y is predicted to increase with a 1-unit increase in X given that Z is held constant. For example, assume that Y is a measure of job performance (supervisory ratings on a 1 = *below expectations* to 7 = *above expectations* scale) and X is a preemployment test of general cognitive abilities (intelligence measured on scale ranging from 1 to 10). Given this illustrative situation, $b_1 = 2$ is interpreted as follows: For

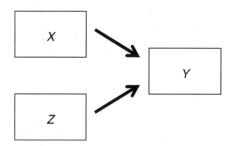

FIGURE 2.1. Graphic representation of Equation 2.1, showing the model including the first-order effects for predicting Y from X and Z.

every 1-point increase in general cognitive abilities, supervisory ratings of performance are predicted to increase 2 points (holding Z constant). As a second illustration, assume now that X is a measure of absenteeism (e.g., number of unexcused absences from work per year). A regression coefficient $b_1 = -.5$ is interpreted as follows: For every unexcused absence from work, supervisory ratings of performance are predicted to decrease half a point (holding Z constant). The interpretation of the intercept a and the regression coefficient b_2 is dictated by the coding scheme that has been used for Z. More detailed information on the interpretation of a and b_2 is provided in Chapters 3 and 8.

The multiple regression model shown in Equation 2.1 assumes that all variables identified by the theory are included in the model and that the variables are properly measured. In addition, it is assumed that the population data have the following characteristics:

1. The relationship between each of the predictors and the criterion is linear.
2. Residuals (i.e., difference between predicted and actual Y scores) exhibit homoscedasticity (i.e., constant variance across values of each predictor; that is, residuals are evenly distributed throughout the regression line).
3. Residuals are independent (i.e., there is no relationship among residuals for any subset of cases in the sample).
4. Residuals are normally distributed.
5. There is less than complete multicollinearity (i.e., perfect correlation between the predictors).

These are the usual assumptions of all ordinary least-squares (OLS) multiple regression models and are described in detail in regression textbooks (e.g., Cohen & Cohen, 1983; Cohen, Cohen, West, & Aiken, 2003; Pedhazur, 1982).

The second equation, called the MMR model, is formed by creating a new variable, the product between the predictors (i.e., $X \cdot Z$), and including it as a third term in the regression. The addition of the product term to the equation yields the following model:

$$Y = a + b_1 X + b_2 Z + b_3 X \cdot Z + e \qquad (2.2)$$

where b_3 is the sample-based least-squares estimate of the population regression coefficient for the product term. This model is represented graphically in Figure 2.2.

Consider the situation where Y is salary after graduation from college, X is number of job offers received, and Z is a hypothesized binary

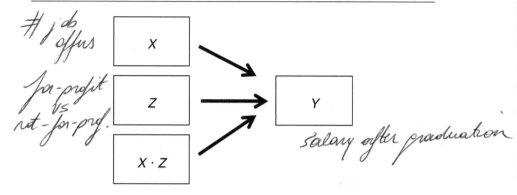

 # j db offus

for-prifit vs. nt-for-prof.

salary after graduation

FIGURE 2.2. Graphic representation of Equation 2.2, showing the model including the first-order effects for predicting Y from X, Z, and the product term $X \cdot Z$ which carries information regarding the moderating effect of Z. This graph is conceptually equivalent to Figure 1.1 (top).

grouping moderator variable called type of industry (i.e., for-profit vs. not-for-profit). In other words, the moderator Z includes two categories or levels, and Z can be coded using one number for members of the for-profit group (i.e., 0) and a different number for members of the not-for-profit group (i.e., 1) (Chapter 8 discusses various coding schemes in detail, including situations when the moderator variable has more than two levels). A researcher forms the hypothesis that the relationship between number of job offers and starting salary is moderated by type of industry, such that the relationship is stronger in for-profit as compared to not-for-profit organizations. The hypothesis is based on the premise that because of the salary compression and fewer resources in most not-for-profit organizations, receiving numerous job offers is not likely to lead to higher starting salaries as compared to receiving fewer offers. On the other hand, for-profit organizations are expected to be able to match offers from other organizations for well-qualified applicants, which is likely to lead to higher starting salaries.

Figure 2.3 shows a scatter plot for the effect of job offers on salary for graduates applying to profit and not-for-profit organizations. This figure seems to provide support for the following hypothesis: There is a stronger relationship for the graduates aspiring to work in for-profit as compared to not-for-profit organizations. Having a larger number of job offers in hand yields greater predicted increases in salary for the graduates applying to work in for-profit as compared to not-for-profit organizations.

To formally test for the moderating effect of type of industry, a

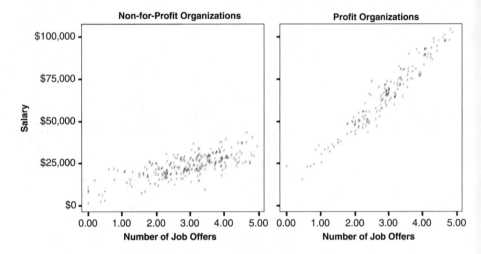

FIGURE 2.3. Hypothetical scatter plots for the relationship between number of job offers and salary for graduates applying to working for-profit (left) and not-for-profit (right) organizations.

"slope"

t statistic can be computed to test the null hypothesis $\beta_3 = 0$. The term β_3 is used to symbolize the regression coefficient for the product term in the population (note that b_3 shown in Equation 2.2 is the product term coefficient in the sample) and should not be confused with beta (i.e., the standardized regression coefficient printed in most computer outputs). Conceptually, this null hypothesis tests whether the amount of change in the slope of the regression of Y on X that results from a unit change in variable Z is greater than would be expected by chance alone.

Alternatively, and equivalently, the coefficients of determination (i.e., squared multiple correlation coefficients, R^2) are compared for Equation 2.1 (i.e., R_1^2) and Equation 2.2 (i.e., R_2^2). The null hypothesis tested is H_0: $\psi_2^2 - \psi_1^2 = 0$, which can also be written as $\Delta\psi^2 = 0$. Conceptually, this null hypothesis tests whether the addition of the product term to the regression equation improves the proportion of explained variance in Y. In other words, this hypothesis answers the question of whether the moderating effect of Z helps improve the prediction of Y above and beyond the first-order effects of X and Z. Note that in some cases R_2^2 may not be statistically significant (Bedeian & Mossholder, 1994). However, the focus is whether the addition of the product term improves the fit of the model predicting Y.

To test H_0: $\psi_2^2 - \psi_1^2 = 0$, an F statistic (distributed with $k_2 - k_1$ and $N - k_2 - 1$ degrees of freedom) is computed using the following formula:

$$F = \frac{(R_2^2 - R_1^2)/(k_2 - k_1)}{(1 - R_2^2)/(N - k_2 - 1)} \tag{2.3}$$

where k_2 is the number of predictors in Equation 2.2, k_1 is the number of predictors in Equation 2.1, and N is the total sample size. Note that the statistical significance levels (i.e., p values) associated with the t and F tests are identical (Cohen & Cohen, 1983).

As a measure of moderating effect size, most researchers choose to focus on ΔR^2, as opposed to b_3, though it is not ideal. The reason for this choice seems to be that ΔR^2 refers to proportion of variance explained and, therefore, is a common metric that can be used to compare effect sizes across studies and areas of research. On the other hand, b_3 is metric specific and is referenced to the specific scales used to measure X, Y, and Z. Consequently, it is difficult to use b_3 to assess the relative size of moderating effects across studies using different scales, even if the studies tested the same moderating-effect hypothesis. A more detailed discussion of ways to assess effect size and the practical importance of moderating effects is provided in Chapter 9.

Furthermore, referring back to Equation 2.2, one may wish to compute a t statistic to test the null hypothesis $\beta_2 = 0$. Note that the term β_2 denotes the regression coefficient for the grouping variable Z in the population, and b_2 is its sample-based least-squares estimate. One may also wish to interpret the meaning of the intercept a. The interpretation of a and b_2 varies, depending on which coding scheme is used. More information on the interpretation of the intercept and coefficients for first-order effects in the presence of an interaction is provided in Chapter 3.

Although no formal tests are performed at this point (a detailed analysis of a fully worked-out example is provided in Chapter 3), Figure 2.3 shows that the slopes seem to differ across groups. Specifically, as hypothesized, there is a steeper (i.e., stronger) slope for the for-profit as compared to the not-for-profit group. Of course, this tentative conclusion based on "eyeballing the data" should be tested analytically by examining the statistical significance of b_3. In addition, Figure 2.3 shows that the intercepts also seem to differ across the groups. Specifically, the regression line crosses the Y-axis at a value of about $9,600 for the for-profit group and at a value of about $8,700 for the not-for-profit group. Again, the conclusion regarding differences between intercepts is tentative until a formal test is conducted. Chapter 3 provides a detailed demonstration of how to implement an MMR analysis using a widely available commercial computer package.

ENDORSEMENT OF MMR
AS AN APPROPRIATE TECHNIQUE

MMR has been recognized as an appropriate technique to assess the presence of moderator variables for half a century (Saunders, 1955, 1956). In spite of the availability of MMR, in the past three decades or so a number of alternative methods have been proposed (Anderson, Stone-Romero, & Tisak, 1996; Arnold, 1982, 1984; Blood & Mullet, 1977; Bobko, 1986; Darrow & Kahl, 1982; Kahl & Darrow, 1984; Morris, Sherman, & Mansfield, 1986). For example, Darrow and Kahl (1982) proposed that the product term be entered before the X and Z predictors in Equation 2.2, based on the argument that this procedure enhances the probability of detecting a moderating effect. Bobko (1986) suggested a set of planned comparisons driven by theory, based on the argument that this procedure increases the power of the moderator test, preserves overall degrees of freedom, and reduces the experiment-wide error rate. Anderson et al. (1996) proposed the use of errors-in-variables regression (EIVR) in lieu of MMR based on the argument that EIVR may not be as adversely affected by measurement error as MMR, thus providing less biased population estimates for the regression coefficients. Hunter and Schmidt (1978) suggested the comparison of X–Y correlation coefficients across groups as a test of the moderating effect. And Morris et al. (1986) proposed the use of principal-components regression (PCR) as an alternative to MMR, based on the argument that PCR is not as adversely affected by multicollinearity (i.e., the correlation between X and Z) as MMR.

Although some of the methods proposed as alternatives to MMR received favorable reviews initially, with the passage of time these methods have been shown to be problematic, and in some cases inappropriate. Stone (1988) and Stone and Hollenbeck (1984, 1989) critically discussed the methods proposed by Arnold (1982, 1984) and Darrow and Kahl (1982). They concluded that MMR was superior, based on theoretical and statistical considerations. Furthermore, the method recommended by Darrow and Kahl (1982; see also Kahl & Darrow, 1984) violates basic principles of multiple regression because the first-order effects must precede (or be entered simultaneously with) the product term in the regression equation; this "backward regression" technique (Stone, 1988) was criticized on the basis of logical and methodological arguments (Stone, 1986; Stone & Hollenbeck, 1984; Tisak, 1994; Wise, Peters, & O'Connor, 1984). The techniques suggested by Bobko (1986) and Morris et al. (1986) were scrutinized by the late Lee Cronbach (1987), who concluded that they are "problematic" and "unacceptable," respectively. Echoing Cronbach's assessment, Stone (1988) criticized

Bobko's (1986) proposed technique, arguing that although comparing a particular cell mean to the mean of the other three cell means in the context of a 2×2 design provides information about a main effect contrast, it does not provide direct evidence regarding the moderating effect. In addition, Dunlap and Kemery (1987) showed that Morris et al.'s (1986) finding of a nonsignificant moderating effect with MMR and significant effects using PCR may have been a result of an artifact of PCR. Hunter and Schmidt's (1978) proposed method to compare correlation coefficients across groups (i.e., "subgroup analysis") considers the correlation coefficients only and does not include a consideration of prediction of criterion scores across moderator-based subgroups. In other words, comparing correlation coefficients (i.e., differential validity) only detracts attention from the more global issue of differential prediction (Bobko & Bartlett, 1978). Instead, using MMR involves a regression equation including means, standard deviations, *and* correlation coefficients (Bobko & Bartlett, 1978). Thus, although they have been treated interchangeably (e.g., Boehm, 1977), differential validity and differential prediction are two related yet distinct issues (Drasgow & Kang, 1984). Finally, Anderson et al.'s (1996) simulation demonstrated that EIVR's parameter estimates were superior to MMR's estimates when both sample size and reliabilities of the predictors were high, but MMR outperformed EIVR in the more common social science situations where reliabilities or sample size are low. In summary, several independent evaluations conducted over the past four decades indicate that MMR is an appropriate method for assessing the effects of moderator variables (Aiken & West, 1991; Cleary, 1968; Cohen & Cohen, 1983; Einhorn & Bass, 1971; Evans, 1991a, 1991b; Fisicaro & Tisak, 1994; Friedrich, 1982; Jaccard et al., 1990; Saunders, 1956; Stone, 1988; Stone & Hollenbeck, 1984, 1989; Stone-Romero & Anderson, 1994; Zedeck, 1971).

Perhaps as a consequence of the numerous independent evaluations concluding that MMR is an appropriate technique for estimating moderating effects, reports issued by several professional organizations of measurement scholars and practitioners have also provided an endorsement for the use of MMR (Sackett & Wilk, 1994). More specifically, such endorsements are found in the *Standards for Educational and Psychological Testing* (American Educational Research Association [AERA], American Psychological Association [APA], & National Council on Measurement in Education [NCME], 1999) and the *Principles for the Validation and Use of Personnel Selection Procedures* (SIOP, 1987). In fact, the latest edition of the *Standards* includes the following statement endorsing the use of MMR (Fairness in Testing and Test Use, Standard 7.6):

> When empirical studies of differential prediction of a criterion for mem-
> bers of different subgroups are conducted, they should include regression
> equations (or an appropriate equivalent) computed separately for each
> group or treatment under consideration or an analysis in which the group
> or treatment variables are entered as moderator variables. (p. 82)

In general, "Under one broadly accepted definition, no bias exists if
the regression equations relating the test and the criterion are indistin-
guishable for the groups in question" (AERA, APA, & NCME, 1999,
p. 79).

In sum, numerous investigations have concluded that MMR is an
appropriate technique to assess the effects of categorical moderator
variables, and MMR has been endorsed by several professional organizations. Consequently, it is rarely surprising that MMR is used so pervasively in the social sciences. But how frequently is MMR used in published research? This question is answered next.

PERVASIVE USE OF MMR IN THE SOCIAL SCIENCES: LITERATURE REVIEW

In some research areas, MMR is clearly the dominant, and often exclusive, methodological tool for assessing moderating effects. One example
is the management accounting literature that focuses on the use of budgets in organizations. In this research area, the key substantive question
is whether the relationship between the use of budgets and a number of
outcome variables (e.g., individual behavioral effects) is contingent on
various moderating organizational variables (Hartmann & Moers,
1999). MMR is used so frequently that it "has become the dominant statistical technique in budgetary research for testing contingency hypotheses" (Hartmann & Moers, 1999, p. 292). A second example is the clinical psychology literature that focuses on aptitude by treatment
interactions resulting in various psychotherapy outcomes (Smith &
Sechrest, 1991). In this literature, the substantive question is whether
the relationship between psychotherapy (i.e., treatment) and psycho-
pathology is contingent on personal characteristics of patients (i.e., aptitude). MMR is also a data analysis tool of choice in this research domain (Smith & Sechrest, 1991). Yet a third illustration of an entire
research area for which MMR is the technique of choice is the job design literature (Champoux & Peters, 1980). In general, job design researchers examine whether the relationship between job design (e.g.,
adding task variety) and various outcomes (e.g., job satisfaction and
work motivation) is moderated by individual and organizational char-

acteristics (e.g., individuals with strong vs. weak growth needs) (Hackman & Oldham, 1976). Champoux and Peters (1980) reviewed 10 years of job design literature and concluded that MMR is the preferred data analysis technique in this research area. Finally, the vast majority of textbooks describing analysis of covariance (ANCOVA) recommend that researchers first check whether the relationship between the criterion and the covariate (i.e., control variable) is similar across groups (e.g., Cohen et al., 2003, pp. 350–351). Thus, before conducting an ANCOVA, researchers are advised to use MMR to test if this important assumption is tenable. Consequently, every researcher who is aware of this critical assumption is likely to use MMR before conducting an ANCOVA.

The above shows how pervasive the use of MMR is in several social sciences. However, in an attempt to quantify the frequency of use of MMR, Aguinis, Beaty, Boik, and Pierce (2003) reviewed all issues of the *Journal of Applied Psychology* (JAP), *Personnel Psychology* (PP), and *Academy of Management Journal* (AMJ) from 1969 to 1998 and identified all articles reporting a discrete moderator analysis test using MMR. Although the review included only three journals, these journals are among the most influential publications devoted to empirical research in applied psychology and management (Starbuck & Mezias, 1996). Thus, they serve as good illustrations of top-tier social science journals devoted to publishing original empirical work.

The criteria for counting a study as using MMR included the following:

1. At least one MMR analysis was included as part of the study.
2. The MMR analysis included a quantitative criterion Y.
3. The MMR analysis included a quantitative predictor X.
4. The MMR analysis included a categorical moderator Z (cf. Equation 2.2).

Aguinis et al. (2003) found 106 articles that fit the preceding criteria. Typically, researchers conducted more than one MMR analysis in any given study. Thus, the total number of reported MMR analyses was 636. Figure 2.4 shows the distribution of MMR analysis over the 30-year period (i.e., 1969–1998) included in the review. As can be seen in this graph, the first article using MMR to assess effects of categorical moderator variables was published in 1977. Also, there is an upward trend in the use of MMR over time. Overall, the frequency of MMR use remained at a high level of approximately 20–40 analyses per year since the mid-1980s.

Given these results based on an admittedly selected set of three

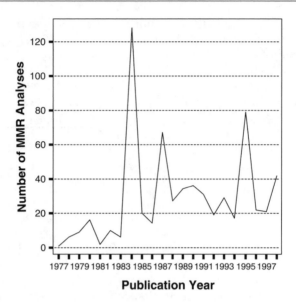

FIGURE 2.4. Number of MMR analyses of categorical moderator variables reported in *Academy of Management Journal, Journal of Applied Psychology,* and *Personnel Psychology* between January 1969 and December 1998. Frequencies for 1969–1976 = 0. Adapted from Aguinis, Beaty, Boik, and Pierce (2003).

journals, one can only guess that the number of MMR analyses reported in all social sciences journals annually is probably in the hundreds. Extrapolating from the Aguinis et al. (2003) review, the number is increasing over time.

CONCLUSIONS

■ This chapter described the MMR model and its basic statistical assumptions. MMR is an extension of a multiple regression equation that includes an additional predictor carrying information regarding the moderating effect. The test of the moderating effect consists of assessing whether the regression coefficient associated with the product term is different from zero in the population; this test is conducted by computing a t statistic. Alternatively, and equivalently, one can assess whether the inclusion of the product term in the regression equation improves our ability to predict the criterion; this test is conducted by computing an F statistic. The statistical significance level of both tests is identical and, therefore, the decision regarding the rejection of the null hypothesis of no moderating effect in the population is identical regardless of which statistical test is used. An examination of the regression coefficient associated with the product term gives us information on whether the Y on X slope differs across moderator-based subgroups.

■ MMR is an appropriate method for estimating moderating effects of categorical variables. Because of the endorsement of using MMR, the procedure is used pervasively in the social sciences. Extrapolating from a selective review of just three journals leads to the conclusion that the number of MMR analyses including categorical variables published annually in social science journals is in the hundreds. This review illustrates that MMR is a method of choice for estimating moderated relationships including categorical moderator variables.

■ The next chapter provides a step-by-step demonstration of how to conduct an MMR analysis using computer programs with an actual illustration using SPSS.

3

Performing and Interpreting Moderated Multiple Regression Analyses Using Computer Programs

I do not fear computers. I fear lack of them.
—ISAAC ASIMOV

Computing is not about computers any more. It is about living.
—NICHOLAS NEGROPONTE

The previous two chapters described the importance of moderator variables for theory and practice, the MMR model, and the pervasive use of MMR in the social sciences. This chapter describes how to conduct an MMR analysis using computer programs and how to interpret the resulting output. Computer programs that conduct MMR analysis have been available for over 30 years (Rock, Barone, & Linn, 1967). This chapter describes the general logic and steps that can be applied in using the commercially available packages (e.g., SAS, SPSS). However, given the widespread use of SPSS, the illustrations in this chapter use SPSS (SPSS, Inc., 1999). This consists of a research scenario involving a moderator variable hypothesis and provides a step-by-step demonstration using a data set available on the Web. To make the best use of this chapter, it would be best to adopt a hands-on approach and read the chapter text while at the same time performing each of the steps on a computer.

RESEARCH SCENARIO *verantwording*

Given the increased calls for accountability in higher education institutions, many universities are implementing systematic and rigorous faculty performance evaluation systems that include measures of teaching, research, and service. Typically, these three measures are combined into an overall performance score that is tied to a specific annual salary increase. Unfortunately, there are budgetary constraints that lead to salary increases that are typically low. Nevertheless, there is an effort to create a more explicit link between faculty performance and monetary rewards.

Consider the situation where we are investigating a newly implemented faculty performance evaluation system. The system was implemented last year, and the research question is whether the relationship between a faculty member's overall performance score (i.e., a combination of research, teaching, and service scores) is a good predictor of his or her salary increase. That is, given the implementation of the new system, is there a clear relationship between faculty performance and pay raises? Also, and more important, we are interested in investigating whether the relationship between performance and salary increase is moderated by tenure status (i.e., untenured vs. tenured). Because of state regulations, the overall performance score is not the only variable that plays a role in affecting salary increases. The deans of the various colleges have the authority to allocate increases in cases where specific individual salaries are substantially below market. Also, state regulations mandate that salary increases be allocated as a percentage of each faculty member's current base salary. Someone with a perfect performance score may have a 5% increase, and someone with an average performance score may have a 3% increase, but these increases are based on the current base salary. Therefore, the salary increases will be larger for the faculty member with the 5% increase, assuming both base salaries are identical. However, a 5% increase may be a smaller dollar figure than a 3% increase if the faculty member with the 3% increase has a higher base salary. Typically, tenured faculty (i.e., associate and full professors) receive higher salaries as compared to untenured faculty (i.e., assistant professors). These discrepancies can be quite large. For example, the 2000–2001 mean annual salary for associate and full professors in accredited business colleges in the United States was $87,250, whereas the mean salary for assistant professors was $73,200 (Association to Advance Collegiate Schools of Business, 2001). Similar differences exist in other social science fields such as psychology (Wicherski, Pate, & Kohout, 2001).

From a theory and past research perspective, a merit-based pay sys-

verdienste

tem increases motivation and is likely to lead to performance improvement if there is a direct link between performance and rewards (Kerr, 1975). However, a merit-based system can lead to perceptions of unfairness and a decrease in motivation and job satisfaction if individuals perceive that this relationship is stronger for one group (e.g., tenured faculty) as compared to another group (e.g., untenured faculty). Given that the base salary is higher for tenured faculty, we expect that tenure status will moderate the relationship between performance and salary increase such that the relationship will be stronger for the tenured group. If such a moderated relationship is found, we would recommend a change in the system.

DATA SET

The data set was generated using the program MULTIVAR (Aguinis, 1994). This program allows for the generation of up to 10 correlated and normally distributed variables, and its executable and source code versions can be downloaded from *http://carbon.cudenver.edu/~haguinis/ mmr/*.

The data set includes 400 cases (i.e., faculty members) and can also be downloaded from *http://carbon.cudenver.edu/~haguinis/mmr/*. The file includes the following three variables:

- Perf: Overall performance score ranging from 1 = *unsatisfactory* to 5 = *exceeds expectations*.
- Salary: Annual salary increase measured in dollars ranging from $13.72 to $2,148.91.
- Tenure: Tenure status, where 0 = *tenured* and 1 = *untenured*. Chapter 8 discusses various coding systems for the categorical moderator variable in detail; for the purpose of this illustration dummy coding is used, such that members of one of the groups are arbitrarily assigned a 0 and members of the other group are assigned a 1. This coding scheme is recommended for situations involving binary moderators because of its simplicity and ease of interpretation of the results.

In short, *Perf* is the predictor, *Salary* is the criterion, and *Tenure* is the hypothesized moderator. Figure 3.1 shows the SPSS data screen.

It is useful to first obtain descriptive statistics to understand the data set better. All statistics computer software packages include procedures to implement these calculations. In SPSS, from the main data screen click on the "Analyze" pull-down menu and choose the "Descriptive Statistics" and "Frequencies" submenus. Double click on each of the three variables to obtain descriptive information on all of them.

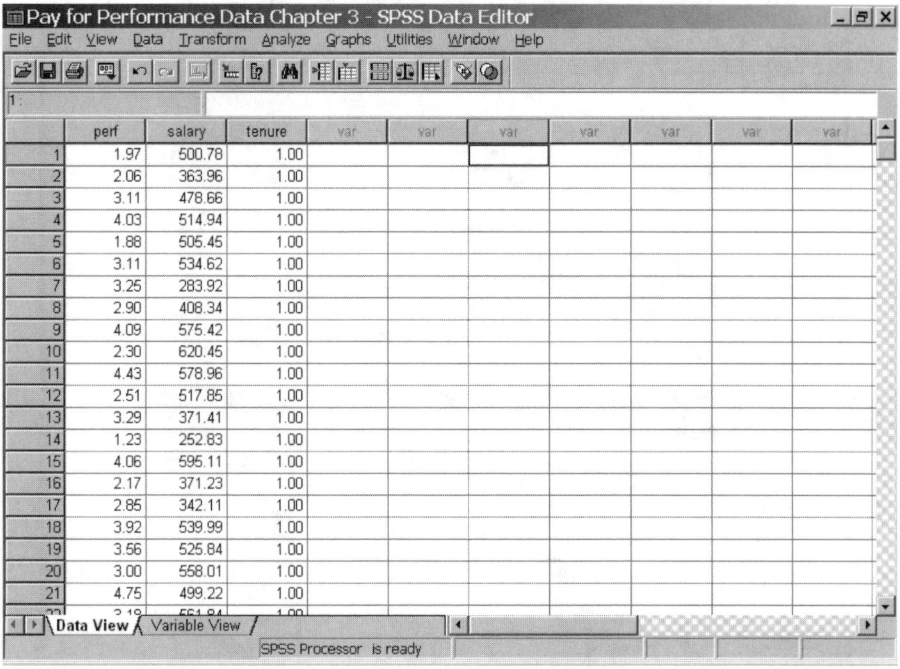

FIGURE 3.1. SPSS data screen for the Pay for Performance data set.

Figure 3.2 (top) shows the resulting screen. Then, click on the "Statistics" option and click on all "Dispersion," "Central Tendency" (except for "Sum," which is not really needed), and "Distribution" choices. Figure 3.2 (bottom) shows the resulting screen.

It is also helpful to obtain some graphs to get a better feel for the data (Tukey, 1977). Thus, from the Frequencies screen, we can click on "Charts" and choose to obtain "Histograms" and also overlap a normal curve (Figure 3.3 shows the resulting screen). Then, we can click on "Continue" and then on "OK."

Results for SPSS's Frequencies procedure are shown in Figure 3.4. Similar output is obtained by using other software packages. This figure confirms that the data set includes 400 cases. Figure 3.4 also shows descriptive statistics for each of the three variables in the data set. Note that the mean for tenure status is .60. Because the coding was such that tenured faculty received a 0 and untenured faculty a 1, a mean of .60 indicates that the sample includes 40% (i.e., 160) tenured individuals. Figure 3.4 also shows that faculty received an annual salary increase ranging from a low of $13.72 to a high of $2,148.91. Overall performance scores ranged from a minimum of 1.00 to a maximum of 5.00.

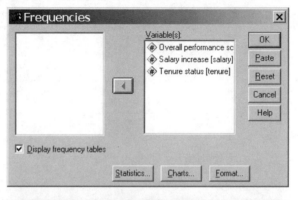

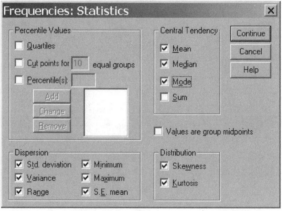

FIGURE 3.2. SPSS screens for the Frequencies procedure (top) and options for Frequencies: Statistics (bottom).

Next is a description of the two steps needed to conduct the MMR analysis and test whether tenure is a moderator of the performance score–salary increase relationship.

CONDUCTING AN MMR ANALYSIS USING COMPUTER PROGRAMS: TWO STEPS

Step 1: Computation of Product Term

Recall that Chapter 2 described that we need to form two regression equations, one including the first-order effects only and a second (i.e., MMR model) including the first-order effects as well as a product term

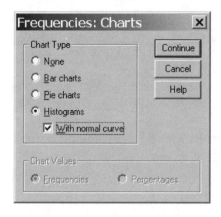

FIGURE 3.3. SPSS screen for Frequencies: Charts.

including the hypothesized moderator variable. In our research scenario, the product term is *Perf × Tenure*. Thus, we first need to create the product term. To do so, we create a new variable consisting of the product between *Perf* and *Tenure*. In SPSS, from the "Transform" pull-down menu, go to the "Compute" option. In the "Target Variable" window,

Output1 - SPSS Viewer

File Edit View Insert Format Analyze Graphs Utilities Window Help

Statistics

		Tenure status	Salary increase	Overall performance score
N	Valid	400	400	400
	Missing	0	0	0
Mean		.6000	817.8175	3.0061
Std. Error of Mean		.02453	25.29688	.04547
Median		1.0000	595.0454	3.0022
Mode		1.00	13.72ª	1.00ª
Std. Deviation		.49051	505.93767	.90944
Variance		.24060	255972.9	.82708
Skewness		-.410	.891	-.038
Std. Error of Skewness		.122	.122	.122
Kurtosis		-1.841	-.352	-.562
Std. Error of Kurtosis		.243	.243	.243
Range		1.00	2135.19	4.00
Minimum		.00	13.72	1.00
Maximum		1.00	2148.91	5.00

a. Multiple modes exist. The smallest value is shown

SPSS Processor is ready

FIGURE 3.4. SPSS output screen for the Frequencies procedure.

enter the name for the product term, for example, "perfxten." Then, in the "Numeric Expression" window, enter "perf * tenure." Figure 3.5 shows the Compute Variable screen once all this information has been entered. Now, click "OK."

The data screen should now include a fourth column with the newly created *perfxten* variable. Figure 3.6 shows the updated data screen.

Step 2: Computation of Regression Equations

As described in Chapter 2, the equations that need to be formed are the following (error terms are omitted for the sake of simplicity):

$$Salary = a + b_1 Perf + b_2 Tenure \tag{3.1}$$

$$Salary = a + b_1 Perf + b_2 Tenure + b_3 Perf \cdot Tenure \tag{3.2}$$

To compute these equations, we need to implement the regression procedure. In SPSS, from the "Analyze" pull-down menu, click on the "Regression" and "Linear" choices. The criterion Salary goes in the "Dependent" variable window, and the predictors *Perf* and *Tenure* go in the "Independent" variable window. Figure 3.7 (top) shows the corresponding SPSS screen.

Note that Figure 3.7 (top) shows that we entered variables *Perf* and *Tenure* in "Block 1 of 1." We now need to enter the product term into the equation to compute Equation 3.2. We do so by clicking on "Next"

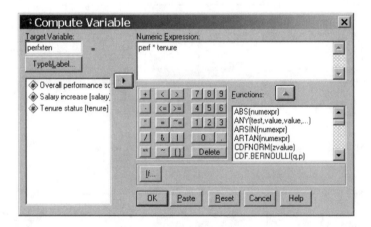

FIGURE 3.5. SPSS screen for creating the product term that carries information regarding the moderating effect of tenure.

```
▦ Pay for Performance Data Chapter 3 - SPSS Data Editor                      _ ⅋ ×
File  Edit  View  Data  Transform  Analyze  Graphs  Utilities  Window  Help
```

	perf	salary	tenure	perfxten	var	var	var	var	var	var
1	1.97	500.78	1.00	1.97						
2	2.06	363.96	1.00	2.06						
3	3.11	478.66	1.00	3.11						
4	4.03	514.94	1.00	4.03						
5	1.88	505.45	1.00	1.88						
6	3.11	534.62	1.00	3.11						
7	3.25	283.92	1.00	3.25						
8	2.90	408.34	1.00	2.90						
9	4.09	575.42	1.00	4.09						
10	2.30	620.45	1.00	2.30						
11	4.43	578.96	1.00	4.43						
12	2.51	517.85	1.00	2.51						
13	3.29	371.41	1.00	3.29						
14	1.23	252.83	1.00	1.23						
15	4.06	595.11	1.00	4.06						
16	2.17	371.23	1.00	2.17						
17	2.85	342.11	1.00	2.85						
18	3.92	539.99	1.00	3.92						
19	3.56	525.84	1.00	3.56						
20	3.00	558.01	1.00	3.00						
21	4.75	499.22	1.00	4.75						
22	3.18	551.94	1.00	3.18						

`◄ ► \ Data View ⋋ Variable View /`   `◄`   `►`

`SPSS Processor  is ready`

FIGURE 3.6. SPSS data screen for the Pay for Performance data set including the newly created *perfxten* variable (i.e., *Perf · Tenure*).

and placing the variable "perfxten" in the Independent variable window. This is shown in Figure 3.7 (bottom).

As noted in Chapter 2, the product term must be entered after its components are already in the equation. If the product term is entered before its components, it is likely to artificially inflate the size of the moderating effect (Stone, 1986, 1988; Tisak, 1994). It is also acceptable to enter all three variables (i.e., X, Z, and the product term) simultaneously in Block 1. However, entering all three predictors as part of Block 1 will not allow the computer program to generate the difference in R^2s between the model with the first-order effects only (i.e., Equation 3.1) and the model with the first-order effects and the product term (i.e., Equation 3.2). Therefore it is advisable to enter the first-order effects first and the product term second.

Finally, once the predictor and the moderator are entered in Block 1 and the product term is entered in Block 2, computer programs allow for several output options. In SPSS, for the "Statistics" options, we can click on estimates and confidence intervals for "Regression Coeffi-

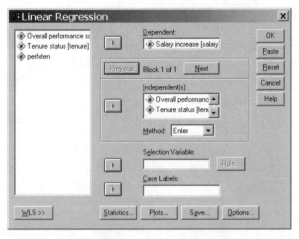

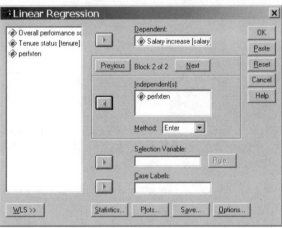

FIGURE 3.7. SPSS screen showing the computation of Equation 3.1 (top) and the addition of variable *perfxten* in Block 2 to compute Equation 3.2 (bottom).

cients," and click on "Model fit," "*R* squared change," and "Descriptives." Figure 3.8 shows this SPSS screen.

Regarding "Plots" options in Figure 3.7, it is useful to produce all standardized residuals plots to check for compliance with the ordinary least-squares assumptions described in Chapter 2 (e.g., homoscedasticity, normality of residuals; see Cohen et al., 2003, pp. 125–141). Then, click on "OK" to run the procedure.

OUTPUT INTERPRETATION

Interpretation of Model 1

All major computer packages include output similar to the information produced by SPSS and shown in Figure 3.9. Figure 3.9 (top) shows that for Model 1 (i.e., Equation 3.1) $R = .925$, $R^2 = .855$, and $F(2, 397) =$ 1168.67, $p = .000$. This R^2 means that 85.5% of the variance in salary in- *92.5%* crease is explained by performance scores and tenure status. Specifically, the Coefficients table from the SPSS output reproduced in Figure 3.9 (bottom) shows that the resulting regression equation for Model 1 is the following:

$$Predicted\ Salary = 669.08 + 222.76\ Perf - 868.14\ Tenure\quad(3.3)$$

Figure 3.9 (top) also indicates that the adjusted $R^2 = .854$. This statistic attempts to correct for capitalization on chance by applying a "correction" factor to R^2 based on the size of the sample and the number of predictors included in the regression model. The smaller the sample and the greater the number of predictors, the smaller the adjusted R^2 (St. John & Roth, 1999). In this illustration, the difference between R^2 and adjusted R^2 is very small because sample size is quite large (i.e., $N = 400$) and the regression equation includes two predictors only.

The coefficients for both performance and tenure in Model 1 are statistically significant at the $p < .001$ level. Equation 3.3 shows that for

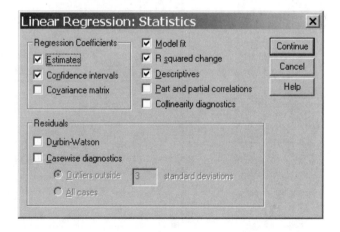

FIGURE 3.8. SPSS screen showing the Linear Regression: Statistics options.

Model Summary

Model	R	R Square	Adjusted R Square	Std. Error of the Estimate	Change Statistics				
					R Square Change	F Change	df1	df2	Sig. F Change
1	.925[a]	.855	.854	193.26707	.855	1168.666	2	397	.000
2	.957[b]	.917	.916	146.72687	.062	292.790	1	396	.000

a. Predictors: (Constant), Tenure status, Overall performance score

b. Predictors: (Constant), Tenure status, Overall performance score, PERFXTEN

Coefficients[a]

Model		Unstandardized Coefficients		Standardized Coefficients	t	Sig.	95% Confidence Interval for B	
		B	Std. Error	Beta			Lower Bound	Upper Bound
1	(Constant)	669.083	35.229		18.992	.000	599.824	738.342
	Overall performance score	222.756	10.641	.400	20.933	.000	201.836	243.676
	Tenure status	-868.138	19.729	-.842	-44.002	.000	-906.926	-829.351
2	(Constant)	183.246	39.006		4.698	.000	106.561	259.932
	Overall performance score	385.620	12.484	.693	30.888	.000	361.077	410.164
	Tenure status	-27.799	51.344	-.027	-.541	.589	-128.740	73.142
	PERFXTEN	-280.196	16.375	-.906	-17.111	.000	-312.389	-248.003

a. Dependent Variable: Salary increase

FIGURE 3.9. SPSS output screen showing Model 1 and Model 2 summary statistics (top) and regression coefficients (bottom).

a 1-point increase in performance score, salary is predicted to increase by $222.76, given that tenure is held constant (i.e., this is the average predicted increase in salary per 1-unit increase in performance across groups). Thus, an improvement in performance of 4 points is likely to lead to an increase of about $900 across groups. This may not be considered a "practically significant" effect for any one given year. However, small salary increases can be meaningful over a 30-year career span (also, a more detailed discussion regarding ways to assess practical significance is provided in Chapter 9). The regression coefficient associated with *Tenure* means that the difference in salary increase between the tenured and untenured groups is $868.14, given that performance score is held constant (i.e., this is the predicted difference in salary between a tenured versus untenured faculty member, assuming their performance scores are equal). Because dummy coding was used and the tenured group received a value of 0, the intercept in Equation 3.3 means that a tenured faculty with an average performance score across the entire sample (i.e., M_{Perf} = 3.0061; see Figure 3.4) is predicted to receive a salary increase of $669.08. Note that the way the data were coded has important implications regarding result interpretation. Specifically, had we used the 0 value for the untenured group and 1 for the tenured group, the intercept would represent the predicted salary increase for an untenured faculty member with an average performance score.

Model 1 does not include the product term and, thus, ignores a possible moderating effect of tenure status. In other words, this model shows that, holding tenure status constant, salary increases by an average of $222.76 when performance increases 1 point. However, so far we have no information regarding the potential moderating effect of tenure on the performance score–salary increase relationship. Could it be that the effect of performance score on salary increase is contingent on tenure status? The answer to this question is given by interpreting Model 2.

Interpretation of Model 2

Model 2 (i.e., Equation 3.2) shows results after the product term has entered the equation. As shown in Figure 3.9 (top), the addition of the product term resulted in an R^2 change of .062, $F(1, 396)$ = 292.79, $p < .001$. This result supports the presence of a moderating effect. In other words, the moderating effect of tenure explains 6.2% of variance in salary increase above and beyond the variance explained by performance scores and tenure status.

The SPSS output shown in Figure 3.9 (bottom) also includes information regarding the regression coefficients after the product term is entered in the equation. The equation is the following:

$$\text{Predicted Salary} = 183.25 + 385.62 \, \textit{Perf} - 27.80 \, \textit{Tenure} - \quad (3.4)$$
$$280.20 \, \textit{Perf} \cdot \textit{Tenure}$$

This output screen also shows that, as described in Chapter 2, the statistical significance (i.e., p value) for the R^2 change from Model 1 to Model 2 based on the F statistic is identical to the statistical significance for the regression coefficient for the product term based on the t statistic (i.e., $p < .001$).

Once again, we base the interpretation of the regression coefficients on the fact that we coded the binary moderator using the dummy coding system. This coding scheme is recommended for situations involving binary moderators because of its simplicity and ease of interpretation of the results. The interpretation of the regression coefficient for the product term in Equation 3.4 is that there is a –$280.20 difference between the slope of salary increase on performance between the untenured (coded as 1) and the tenured group (coded as 0). In other words, the slope regressing salary on performance is less steep for untenured faculty members as compared to tenured faculty members.

Because the interpretation of the coefficient of the product term can be confusing, it is typically useful to create a graph displaying the performance–salary relationship for each of the groups. To do so, we first need to construct the regression equation for each of the two groups. Recall that tenured faculty were assigned a code of 0, whereas untenured faculty were assigned a code of 1. Therefore, reworking Equation 3.4 for the tenured group (i.e., $\textit{Tenure} = 0$) yields the following:

$$\text{Predicted Salary} = 183.25 + 385.62 \, \textit{Perf} - 27.80 \, \textit{Tenure} - \quad (3.5)$$
$$280.20 \, \textit{Perf} \cdot \textit{Tenure}$$

$$\text{Predicted Salary} = 183.25 + 385.62 \, \textit{Perf} - 27.80(0) - 280.20(0)$$

Tenured Faculty: $\text{Predicted Salary} = 183.25 + 385.62 \, \textit{Perf}$

Reworking Equation 3.4 for the untenured group (i.e., $\textit{Tenure} = 1$) yields the following:

$$\text{Predicted Salary} = 183.25 + 385.62 \, \textit{Perf} - 27.80 \, \textit{Tenure} - \quad (3.6)$$
$$280.20 \, \textit{Perf} \cdot \textit{Tenure}$$

$$\text{Predicted Salary} = 183.25 + 385.62 \, \textit{Perf} - 27.80(1) - 280.20 \, \textit{Perf}(1)$$

Predicted Salary = 183.25 + 385.62 *Perf* − 27.80 − 280.20 *Perf*

Predicted Salary = 155.45 + *Perf* (385.62 − 280.20)

Untenured Faculty: *Predicted Salary* = 155.45 + 105.42 *Perf*

Now, we can plot the performance–salary relationship for each group. To do so, it is recommended that we choose values of 1 standard deviation (*SD*) above and below the mean for *Performance* in Equations 3.5 and 3.6 (Cohen et al., 2003). Figure 3.4 shows that the mean score for *Performance* is 3.00, and the *SD* is .91. So, using the value of 3.91 (1 *SD* above the mean) and 2.09 (1 *SD* below the mean) in Equations 3.5 and 3.6 yields the graph shown in Figure 3.10.

Results based on Equation 3.4 led to the conclusion that there is a moderating effect of tenure. A perusal of Figure 3.10 showing the performance score–salary increase relationship for each of the groups separately gives us a better sense that the relationship is stronger (i.e., steeper slope) for the tenured faculty as compared to the untenured faculty group.

Additional Issues in Output Interpretation

Results indicate a moderating effect such that the relationship between performance and salary increase is stronger for tenured as compared to untenured faculty members. In some situations, it may be useful to interpret the coefficient associated with performance score, ignoring the moderating effect (i.e., Model 1). Because we are not taking into account the interaction effect, this coefficient can be considered an *average effect* of the relationship between performance and

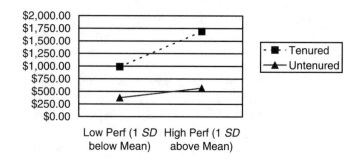

FIGURE 3.10. Slopes for *Salary* on *Performance* for tenured and untenured faculty based on Equations 3.5 and 3.6.

salary across levels of the moderator variable (i.e., tenure status) (Jaccard et al., 1990; Overall, Lee, & Hornick, 1981). The presence of the interaction implies that this average was computed from heterogeneous values (i.e., a larger coefficient for tenured than untenured faculty). Therefore, the interpretation of the interaction yields more detailed and precise information about the relationship between performance and salary for various groups (i.e., tenured vs. untenured faculty). On the other hand, the interpretation of the first-order effects provides less precise information.

Additional considerations are needed if we wish to interpret the intercept and the coefficient associated with the first-order effect of the predictor hypothesized to be the moderator in the presence of a non-zero moderating effect in Model 2 (Katrichis, 1993). Consider Equation 3.4, including the moderating effect of tenure. The interpretation of the regression coefficient for *Tenure* is that the estimated difference between the salary increase of an untenured faculty member and a tenured one, both with a performance score of 0, is –$27.80. But, a score of 0 does not even exist for *Performance* because this variable was measured on a scale ranging from 1 to 5! Thus, the regression coefficient for *Tenure* in Equation 3.4 is not really meaningful. Similarly, the intercept in the model including the product term is interpreted as the salary increase for a member of the tenured group (because tenure was coded as 0) with a performance score of 0. Again, this is not very meaningful because the value of 0 falls outside of the 1–5 scale used to measure performance. Note, however, that the coefficient for the moderator and the intercept are more readily interpretable when the predictor X scale includes a meaningful zero point.

One way to make the first-order coefficient for the moderator and the intercept more interpretable is to center the quantitative predictors (Aiken & West, 1991, Chapter 3). Centering achieves the goal of making zero a meaningful value and, consequently, the coefficients become easier to understand. Centered scores are obtained by simply subtracting the mean from each score, resulting in transformed scores with a mean = 0 (Tate, 1984). It should be noted, however, that in some situations the MMR models based on centered and uncentered predictors may be functionally equivalent (Kromrey & Foster-Johnson, 1998). Also, centering is just one of several possible ways of making the zero point more meaningful. An alternative procedure includes transforming the zero point into the scale midpoint or a neutral midpoint; for example, $Performance_{mid} = Performance - 3$ (because in this case 3 is the midpoint on the scale used to measure *Performance*). Another alternative is to transform the zero point into the median: $Performance_{med} = Performance - MDN_{Performance}$. In the latter case, the coefficient associated with

Tenure is interpreted as the effect of *Tenure* on *Performance* at the median score for *Performance*.

A simple way to perform the centering procedure using computer packages is to create a new "centered performance" variable. In SPSS, this is achieved by using the "Compute" command. Using our data set, the centered variable is obtained as follows: *mean performance* ↓

$$\text{Centered Performance } [cperf] = Perf - 3.0061 \qquad (3.7)$$

As a way to make sure the procedure has been conducted in the appropriate way, it is a good idea to run the Frequencies procedure and obtain descriptive statistics for the centered variables. If centering was done correctly, the mean score for *cperf* should be 0. Then the centered variables and the product between the centered variables are used to re-run the MMR analyses. The resulting MMR model is the following:

$$\text{Predicted Salary} = 1342.46 + 385.62 \ cperf - \qquad (3.8)$$
$$870.10 \ tenure - 280.20 \ cperf \cdot tenure$$

= 0

Based on Equation 3.8, the intercept means that the estimated salary increase for a tenured faculty member with an average performance score (i.e., 3.00) is $1,342.46, taking into account the performance–salary increase relationship for the tenured group. The regression coefficient associated with tenure means that the difference *inter-* in salary increase between an untenured and a tenured faculty mem- *action* ber with average performance scores is –$870.10. The regression coef- *term =* ficient associated with centered performance means that the salary in- *0* crease for a tenured faculty member with an average performance *nice* score is $385.62. This is identical to the slope that would be obtained if we regressed salary increase on performance for the group of ten- *cperf=0* ured faculty only (i.e., for tenured faculty *Predicted Salary* = 183.25 + 385.62 *Perf*). Again, the interpretation of these coefficients is guided by the fact that dummy coding was used. *4? = 1342.46?*

A comparison of Equation 3.4, using uncentered performance scores, with Equation 3.8, using centered performance scores, reveals that there are important changes in the intercept as well as the coefficient associated with the effect of *Tenure*. However, the coefficients associated with the quantitative predictor and the product term have remained unchanged (Lautenschlager & Mendoza, 1986). Also, the R^2s associated with Model 1 and Model 2 are identical to the situation including uncentered predictors. This is because simple additive transformations on predictor variable scores such as centering do not change the statistical test of the product term (Arnold & Evans, 1979; Cohen,

1978; Friedrich, 1982). Similarly, the R^2s associated with Model 1 and Model 2 are not affected by the use of different coding schemes.

Note that there is no need to center the quantitative criterion (i.e., salary increase, in this example). If the criterion is in its original uncentered scale, then predicted scores are also expressed in the same units as the original scale. Thus, the interpretation of predicted criterion scores is more straightforward when left in their original metric.

Transformations involving adding or subtracting a constant such as centering also do not affect regression coefficients in Model 1 (i.e., Equation 3.3) (but they may affect the intercept). As noted earlier, they do not affect the coefficient for the product term in Model 2 (i.e., Equation 3.4) (Irwin & McClelland, 2001). However, as seen earlier, these types of transformations have dramatic effects on the coefficients of the lower-order terms because coefficients resulting from uncentered variables are referenced to a zero point on the other predictor, whereas coefficients resulting from centered variables are referenced to the *average* value on the other predictor. Nevertheless, the interpretation of the moderating effect is identical in equations based on centered and uncentered predictors. Consequently, as long as the scales have interval-level properties and arbitrary zero points (i.e., Likert-type scales) (Aguinis, Henle, & Ostroff, 2001), it is still appropriate to test moderating effects.

It should be emphasized that interpretation of the intercept and the regression coefficient associated with the moderator is likely to change if a different coding scheme is used (Schoorman, Bobko, & Rentsch, 1991). Because each coding scheme represents a different way to partition the total variance associated with the interaction, the regression coefficients associated with the first-order effects and the intercept answer a different question for each coding scheme and, therefore, their value and statistical significance level also vary (Schoorman et al., 1991). However, the value and significance level for the coefficient for the product term as well as the model's R^2 are not affected by the use of a different coding type. A more detailed discussion of the effects of various coding strategies on the interpretation of the results is provided in Chapter 8.

Finally, in interpreting the computer output for Model 2, MMR researchers are better served by examining the unstandardized regression coefficients (i.e., bs) as opposed to the standardized coefficients (i.e., betas) (Friedrich, 1982; Jaccard et al., 1990). The interpretation of the "standardized" solution provided by computer packages is fraught with great difficulties because even if the predictors are standardized, their product is not necessarily standardized. This is because instead of first standardizing X and Z (i.e., Z_X, Z_Z) and then obtaining their product (i.e., $Z_X \cdot Z_Z$), computer programs first generate the product between X and Z and then standardize the resulting scores (i.e., $Z_{X \cdot Z}$). Because

$Z_X \cdot Z_Z$ and $Z_{X \cdot Z}$ may differ, the "standardized" solution produced by computer outputs may be misleading. If an MMR user is interested in obtaining regression coefficients that are interpretable in a standardized metric, this is possible by first converting each of the predictors and the criterion into standard scores (i.e., [score − mean]/SD), and then creating the product term based on the standard scores for X and Z (i.e., $Z_X \cdot Z_Z$). Then, because all variables have been converted to standard scores, the bs shown in the computer output can be interpreted using a standard score metric (Friedrich, 1982; Jaccard et al., 1990).

Figure 3.11 shows the SPSS output screen corresponding to the MMR analysis including standardized scores for *Salary*, *Performance*, and *Tenure*. For Model 2, the interpretation of the unstandardized coefficient associated with *Standardized Tenure* is that the difference in standardized salary increase between an untenured and a tenured faculty member with average performance scores is .844. The interpretation of the unstandardized coefficient for *Standardized Performance* is that the mean standardized salary increase for a tenured faculty member with an average performance score is .39. Finally, the unstandardized coefficient associated with the product terms means that the standardized difference for the slope of salary increase on performance between the untenured (coded as 1) and the tenured group (coded as 0) is −.247. In other words, the slope is steeper for the tenured group by .247 salary-increase standard deviation units.

CONCLUSIONS

■ This chapter provided a step-by-step description of how to conduct an MMR analysis including a binary moderator variable using computer programs, with a special emphasis on SPSS. The procedure involves creating a new variable that consists of the product term between the predictor and the moderator variables, and implementing a hierarchical regression procedure. This chapter described how to interpret the computer output, including answering the key issue of whether the Y on X slope differs across the moderator-based subgroups.

■ The chapter used dummy coding for the binary moderator. In this coding scheme, members of one of the groups are arbitrarily assigned a 0 and members of the other group are assigned a 1. This coding scheme is recommended for situations involving binary moderators because of its simplicity and ease of interpretation of the results. The choice of a coding scheme affects the interpretation of the intercept and the regression coefficient associated with the first-

Coefficients[a]

Model		Unstandardized Coefficients		Standardized Coefficients	t	Sig.	95% Confidence Interval for B	
		B	Std. Error	Beta			Lower Bound	Upper Bound
1	(Constant)	1.860E-05	.019		.001	.999	-.038	.038
	ZPERF	.400	.019	.400	20.933	.000	.363	.438
	ZTENURE	-.842	.019	-.842	-44.002	.000	-.879	-.804
2	(Constant)	5.110E-03	.015		.352	.725	-.023	.034
	ZPERF	.391	.015	.391	26.904	.000	.362	.420
	ZTENURE	-.844	.015	-.844	-58.088	.000	-.872	-.815
	ZPRODUCT	-.247	.014	-.249	-17.111	.000	-.275	-.219

a. Dependent Variable: ZSALARY

FIGURE 3.11. SPSS output screen showing Model 1 and Model 2 regression coefficients based on standardized scores.

order effect of the moderator in the MMR model. Coding schemes other than dummy coding represent a different way of partitioning the total variance associated with the interaction. Therefore, the regression coefficients associated with the first-order effects and the intercept answer a different question for each coding scheme and, consequently, their value and statistical significance level also vary. On the other hand, the value and significance level for the coefficient for the product term as well as the model's R^2 are not affected by a change in the coding scheme.

- Centering of the predictor X is an additional issue that should be taken into account if a researcher wishes to interpret the first-order coefficients in the presence of a nonzero interaction. The coefficient for the first-order effect of the moderator variable is referenced to a *zero* point on the other predictor, and this "zero point" may not be meaningful given the scale used to measure the predictor X. On the other hand, the coefficient resulting from centering the X variable is referenced to the *average* value on X. Thus, implementing an additive transformation on X such as centering is likely to result in a more meaningful coefficient for the first-order effect of the moderator. Note, however, that the size, statistical significance, and interpretation of the coefficient associated with the product term are identical in equations based on centered and uncentered predictors.

- Going back to the research questions posed at the beginning of this chapter, the finding is that there is an overall positive relationship between performance scores and salary increase. Regarding the substantive moderating-effect hypothesis, tenure status is a moderator such that the relationship is stronger for tenured than for untenured faculty. In practical terms, untenured faculty do not receive as high a pay increase as compared to tenured faculty given the same performance score. This is a critical piece of information in implementing the new performance management system. It is likely that untenured faculty will perceive the system to be unfair. Therefore, the system may have an effect exactly opposite to what was intended: The new performance management system may serve as a factor that decreases the motivation and satisfaction of untenured faculty.

- The next chapter addresses a critical statistical assumption that, if violated, may bias conclusions regarding the presence of a moderating effect. Specifically, Chapter 4 addresses a fundamental yet often ignored assumption of MMR: Homogeneity of error variance.

4

The Homogeneity of Error Variance Assumption

A statistician is a person who draws a mathematically precise line from an unwarranted assumption to a foregone conclusion.
—ANONYMOUS

This chapter addresses an important issue that may adversely affect conclusions regarding the presence of a moderator. Specifically, this chapter addresses a fundamental yet often ignored assumption of MMR: homogeneity of error variance. When this assumption is violated, heterogeneity of error variance exists and the conclusions of the moderator analysis may be incorrect. This chapter describes this assumption, differentiates it from the homoscedasticity assumption, and includes the findings of a review of relevant literature, which indicates, unfortunately, that the assumption is violated quite frequently in published research. The chapter also shows the effects of violating the homogeneity of error variance assumption on research conclusions based on MMR analysis. Finally, it describes a computer program available on the Web that allows researchers to (1) check whether the assumption has been violated and (2) conduct a test for moderation using analyses other than MMR when there is a violation of the assumption.

WHAT IS THE HOMOGENEITY OF ERROR VARIANCE ASSUMPTION?

Chapter 2 describes the assumptions of ordinary least-squares (OLS) regression models. Specifically, the MMR model assumes that the variables identified in the theoretical construct are all included in the equation and that the variables are properly measured. In addition, it is assumed that population data have the following characteristics:

1. The relationship between each of the predictors and the criterion is linear.
2. Residuals (i.e., difference between predicted and actual Y scores) exhibit homoscedasticity (i.e., constant variance across values of each predictor; in other words, residuals are evenly distributed throughout the regression line).
3. Residuals are independent (i.e., there is no relationship among residuals for any subset of cases in the sample).
4. Residuals are normally distributed.
5. There is less than complete multicollinearity (i.e., perfect correlation between the predictors).

These are the usual assumptions of all OLS multiple regression models and are described in detail in regression textbooks (e.g., Cohen & Cohen, 1983; Cohen et al., 2003; Pedhazur, 1982).

When MMR is used to assess the potential effects of categorical moderator variables, a critical assumption is that the variance in Y that remains after predicting Y from X is equal across moderator-based subgroups. This is called the homogeneity of (within-group) error variance assumption. The assumption implies that the error variance for one moderator-based population (i.e., $\sigma^2_{e_1}$ or $\sigma^2_{Y_1 - \hat{Y}_1}$) is equal to the error variance in each of the other moderator-based populations (e.g., $\sigma^2_{e_i}$ or $\sigma^2_{Y_i - \hat{Y}_i}$). In other words, the predicted scores for Y should be similarly distributed about the regression line for each of the moderator-based populations. In the context of analysis of variance (ANOVA) models, $Y - \hat{Y}$ scores are labeled residual scores. Thus, the homogeneity of error variance assumption in MMR is similar to the homogeneity of variance assumption in the context of ANOVA.

In symbols, the error variance for each of the moderator-based populations is

perfect correlation $\sigma^2_{e_i} = 0$

$$\sigma^2_{e_i} = \sigma^2_{Y_i}(1 - \rho^2_{XY_i})$$

zero correlation $\sigma^2_{e_i} = \sigma^2_{Y_i}$ (4.1)

where $\sigma^2_{Y_i}$ and $\rho^2_{XY_i}$ are the Y variance and the X–Y correlation in each moderator-based population, respectively.

Because MMR users do not typically have access to population values, $\sigma^2_{e_i}$ can be estimated by computing $s^2_{e_i}$ as follows:

$\rho = \rho\varphi,\ z =$ *Sample*

$$s^2_{e_i} = \frac{s^2_{Y_i}(n_i - 1)(1 - r^2_{XY_i})}{n_i - 2}$$ (4.2)

Subgroups sample size

where n_i is the subgroup sample size. Note that as subgroup sample size increases, Equations 4.1 and 4.2 provide similar results.

if all random variables have the same finite variance

TWO DISTINCT ASSUMPTIONS: HOMOSCEDASTICITY AND HOMOGENEITY OF ERROR VARIANCE

similarity

Homoscedasticity (also labeled Type I homoscedasticity) (Wilcox, 1997b) applies to all ordinary least-squares regression models (including MMR), whereas the homogeneity of error variance assumption (also labeled Type II homoscedasticity; Wilcox, 1997b) applies only to MMR models including categorical moderators. Fundamentally, both assumptions refer to the same issue: a desired constant distribution of residual scores around a regression line. However, homoscedasticity refers to whether there is a constant distribution of residuals for a set of individual scores, whereas homogeneity of error variance refers to whether the distribution of residuals is constant across the moderator-based categories. MMR users should not presume that meeting the homoscedasticity assumption implies that the homogeneity of error variance assumption is also satisfied. In the presence of homoscedasticity, the homogeneity of error variance assumption may or may not be satisfied.

It may be useful to illustrate the distinction between the assumptions with a research scenario. Consider the situation where a school psychologist has developed a scholastic aptitude test aimed at predicting scholastic achievement (i.e., grade-point average [GPA]) for middle school children. The test will be used as part of a decision-making process that results in student placement in various levels. Because the results of the assessment will be applied in an ethnically diverse school district, the psychologist wishes to determine if the relationship between test scores and GPA is moderated by ethnicity (i.e., Whites vs. ethnic minorities). That is, the psychologist is interested in the potential moderating effect of ethnicity.

The school psychologist can check whether the homoscedasticity assumption is satisfied for the regression model including both White and ethnic minority students (we label this "overall homoscedasticity"). To do this, the researcher needs to examine the GPA on a test scores regression model including all scores (i.e., Whites and ethnic minorities combined). The homoscedasticity assumption is satisfied if the residual scores are similarly distributed across various points of the X scale.

As was noted earlier, it is possible to comply with the homoscedasticity assumption (i.e., Type I homoscedasticity) and yet violate the homogeneity of error variance assumption (i.e., Type II homoscedasticity). It is also possible to have complete heteroscedasticity (i.e., Type I and Type II heteroscedasticity). In fact, it is possible to have overall homoscedasticity and homoscedasticity in each of the moderator-based subpopulations, and yet violate the homogeneity of error variance assumption. For example, if the Y variances are equal across the two populations (i.e., $\sigma_{Y_1}^2 = \sigma_{Y_2}^2$) and there is a stronger relationship between test scores and achievement scores for the population of White students than the ethnic minority students (i.e., $\rho_{XY_1} > \rho_{XY_2}$), Equation 4.1 indicates that the error variances *must* differ across subpopulations (i.e., $\sigma_{e_1}^2 < \sigma_{e_2}^2$). That is, if ρ_{XY}s differ across populations, σ_e^2s must differ as well unless the difference in Y variances is precisely counterbalanced by the difference in correlation coefficients. Similarly, if the correlation coefficients are equal across subpopulations, but Y variances differ, this situation also leads to a systematic violation of the homogeneity of error variance assumption, even in the presence of subpopulation and overall homoscedasticity. Finally, both the Y variances and correlations may differ across subpopulations, which also results in the violation of the assumption.

It is helpful to complement the preceding analytic discussion with the use of graphs. Figure 4.1 shows a scatter plot of the relationship between test scores and GPA for the entire sample of 250 students. As can be seen in this figure, predicted GPA scores are evenly distributed throughout the regression line. Thus, the homoscedasticity assumption has been met in the entire sample (i.e., Type I homoscedasticity).

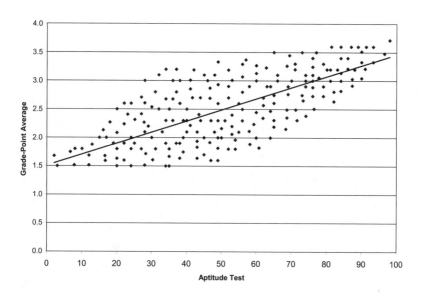

FIGURE 4.1. Scatter plot of a hypothetical relationship between aptitude test scores and scholastic achievement for all students combined.

Now, we can create two separate scatter plots, each one showing the test scores–GPA relationship for each of our two groups under consideration (Whites and ethnic minorities). Figure 4.2 shows the scatter plot for Whites ($n = 153$), and Figure 4.3 shows the scatter plot for ethnic minorities ($n = 87$). As in the case of the complete sample, predicted GPA scores are evenly distributed along the regression line for each of the two groups. Thus, the homoscedasticity assumption has been met for each of the two groups.

In spite of the presence of overall homoscedasticity and homoscedasticity in each of the two groups, Figures 4.2 and 4.3 show that the amount of error variance present when GPA is predicted from test scores is clearly not equivalent for the two moderator-based subgroups (i.e., Type II heteroscedasticity). Although the data points are evenly distributed throughout each of the two regression lines, in this example the average deviation of the data points from the line is larger for ethnic minority students (Figure 4.2) than for White students (Figure 4.1). This can be seen because the scatter cloud is wider for the sample of ethnic minority students as compared to the sample of White students. Stated differently, the amount of error variance (s_e^2) is not equivalent across the two moderator-based subgroups; that is, it is larger for the ethnic minority (Figure 4.3) than the White (Figure 4.2) subgroup (i.e., $s_{e_1}^2 < s_{e_2}^2$).

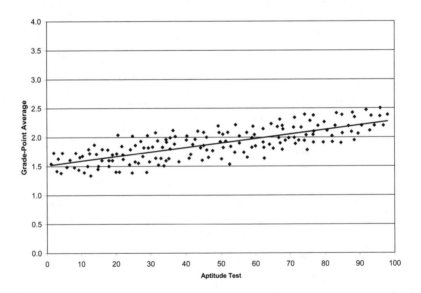

FIGURE 4.2. Scatter plot of a hypothetical relationship between aptitude test scores and scholastic achievement shown in Figure 4.1, for White students only.

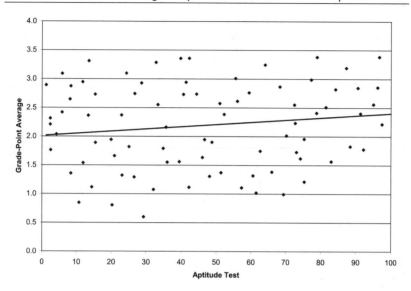

FIGURE 4.3. Scatter plot of a hypothetical relationship between aptitude test scores and scholastic achievement shown in Figure 4.1, for ethnic minority students only.

IS IT A BIG DEAL TO VIOLATE THE ASSUMPTION?

So far, this chapter has described the homogeneity of error variance assumption (i.e., Type II homoscedasticity) and clarified that it is different from the homoscedasticity assumption (i.e., Type I homoscedasticity). Unfortunately, many MMR users are not aware of the homogeneity of error variance assumption, or maybe they are aware of it and choose to ignore it. Perhaps a reason for the lack of attention given to this assumption is the perception that the assumption is not frequently violated and, if it is, violating the assumption may not be perceived as having important consequences for moderator analysis (Oswald, Saad, & Sackett, 2000). This lack of attention exists in spite of the recommendation that because the MMR test "assumes equal population within-group residual variances, inspection of the assumption of a common σ^2 should precede the test of equal slopes" (Rogosa, 1980, p. 310).

Aguinis, Petersen, and Pierce (1999)—in a different review from the one described in Chapter 2 (Aguinis, Beaty, Boik, & Pierce, 2003), but that reviews the same journals—conducted a review of articles published from January 1987 to April 1999 in *Academy of Management Journal, Journal of Applied Psychology,* and *Personnel Psychology.* These three journals were selected because they are among the most influential and methodologically rigorous publications devoted to empirical research in applied psychology (Starbuck & Mezias, 1996). Aguinis et al. (1999) found that 87 articles published in these journals during the 1987–1999 period used MMR to assess effects of categorical moderators. But, only one (i.e., Stewart, Carson, & Cardy, 1996) reported having conducted an assessment of compliance with the homogeneity of error variance assumption. It is likely that MMR users do not pay attention to this assumption because they may not be aware of the consequences of this violation. But, what are the consequences of violating the assumption? Could it be that MMR is robust to violating this assumption? In other words, should we really care about complying with this assumption before conducting an MMR analysis?

Obviously, the discussion of this assumption would not have such a prominent place in this book if it were not important. Thus, the answer to the preceding questions is that it *does* matter if the assumption is violated. In fact, violating the assumption can affect MMR results and conclusions substantially. In the presence of heterogeneity of error variance, researchers may commit a Type I or a Type II statistical error, depending on the specific sample and population characteristics. Thus, users of MMR may discover a *false* moderator (Type I error) or *erroneously* dismiss a model including a moderator variable (Type II error).

Returning to the situation faced by the school psychologist, a Type I error means that he or she may conclude the test is biased toward ethnic minorities, but in actuality the test is not. Thus, the school psychologist may make the incorrect decision to spend more time and effort to develop additional items that may decrease the (false) bias, or choose to not use the test altogether. Obviously, this represents an unfortunate waste of time and resources that went into developing a test that may actually be unbiased toward ethnic minority students. On the other hand, making a Type II error means that the school psychologist concludes the test is not biased, but in actuality the test is biased. Accordingly, he or she may decide to go ahead and use the test to predict student achievement and make decisions about student placement in various levels. Because MMR analysis led to the conclusion that the test is not biased, the researcher is not aware that decision making about student placement is erroneous and may be penalizing ethnic minority students unfairly. Aguinis and Pierce (1998a) reviewed the effects of violating the assumption on Type I and Type II errors in more detail. The following is a summary of their review.

Violation of the Homogeneity of Error Variance Assumption: Type I Error Rates

When is a violation of the homogeneity assumption likely to lead to a Type I error (i.e., find a false moderator)? Dretzke, Levin, and Serlin

(1982) and DeShon and Alexander (1996) presented results of computer simulations that examined the effects of violating the homogeneity of error variance assumption on Type I error rates. These simulations consisted of randomly selecting samples from populations of scores with known characteristics. It was known, for instance, that there was no moderating effect and the degree to which the homogeneity of error variance assumption was violated was also known. Then Dretzke et al. (1982) and DeShon and Alexander (1996) conducted MMR analyses on samples drawn from these populations and assessed whether the sample-based conclusions were correct or incorrect vis-à-vis the known characteristics of the populations. In this case, because populations were created such that there was no moderating effect, researchers investigated whether Type I error rates remained close to the preset nominal level (typically .05) when the assumption was violated.

Dretzke et al. (1982) found that when sample sizes are equal across moderator-based subgroups, Type I error rates associated with a null hypothesis of no moderating effect do not seem to be artificially inflated. Alternatively, in the more typical situation of unequal subgroup sample sizes (e.g., more men than women, more Whites than ethnic minorities), error variance heterogeneity can result in an inflated Type I error rate when testing for moderating effects. For example, when sample size (n) is 50 and the correlation (r) between the quantitative predictor and the criterion is .25 in one moderator-based subgroup and n is 100 and r is .75 in the second subgroup, the actual Type I error probability using MMR was .18 when the preset nominal Type I error was set at .05 (Dretzke et al., 1982).

Dretzke et al. (1982) discovered that Type I error rate inflation was most noticeable when the smaller subgroup sample size was paired with the larger residual variance. Dretzke et al.'s (1982) simulation held the variance of the quantitative predictor X constant across subgroups but did not hold constant the variance of the criterion Y across subgroups (this is why subgroup correlations differed; cf. Equation 4.1). Subsequently, DeShon and Alexander (1996) showed that Dretzke et al.'s results that MMR results were not affected by violation of the assumption when sample size was equal across groups should be qualified. DeShon and Alexander's simulation indicated that Dretzke et al.'s conclusions hold only when the X variance is equal across subgroups. However, when the X variance is moderately unequal across subgroups, and sample sizes are equal, heterogeneity of error variance leads to overly conservative Type I error rates.

In conclusion, what happens when there is no moderating effect in the population and MMR users are faced with heterogeneity of error variance? Aguinis and Pierce's (1998a) literature review noted the following two conclusions. First, Type I error rates are likely to be artificially inflated when sample sizes are *unequal* across subgroups. This is most noticeable when the smaller subgroup sample size is paired with the larger error variance. Second, Type I error rates are also affected under conditions of *equal* subgroup sample sizes. Type I error rates become overly conservative when the X variance is dissimilar across subgroups. Thus, MMR users should be especially aware that heterogeneity of error variance can cause inaccurate Type I error rates when (1) sample sizes are unequal across subgroups (resulting in overly liberal Type I error rates), and (2) sample sizes are equal across subgroups and X variances are unequal across subgroups (resulting in overly conservative Type I error rates).

Violation of the Homogeneity of Error Variance Assumption: Type II Error Rates

When is a violation of the homogeneity assumption likely to lead to a Type II error (i.e., miss an existing moderating effect)? Computer simulations addressing the effects of violating the assumption on Type II error rates proceed as follows. First, populations are created in which there is a moderating effect. Then, samples are drawn from these populations of scores under various conditions ranging from a mild violation of the assumption (i.e., error variances are only slightly different across groups) to a very severe violation of the assumption (i.e., error variances differ substantially across groups). Next, MMR analyses are conducted on samples drawn from these populations and results are assessed regarding the number of samples in which the moderating effect was not detected (i.e., Type II error rate). Typically, MMR users are more interested in statistical power than Type II error rates. Statistical power equals 1 – Type II error rate, and it is the probability that an MMR analysis will reveal the presence of a moderator when there is one in the population.

Alexander and DeShon (1994) conducted a Monte Carlo study to investigate the impact of violating the homogeneity of error assumption on statistical power. A Monte Carlo study involves computer simulations that allow researchers to understand the probabilities of different possible outcomes (e.g., MMR's Type I error rates and statistical power). They found that when sample sizes are unequal across groups and the group with the larger sample size is associated with the larger error variance, violating the assumption causes statistical power to decrease substantially. For example, in the case with two groups and group ns (rs) of 20 (.20) and 40 (.50), the statistical power of the MMR test was 1.00 (i.e., MMR results based on each sample resulted in the correct conclusion that there is a moderator in every case). However, when the group

n, sample size   correlation

with the larger sample size was paired with the smaller correlation (i.e., when ns [rs] were 20 [.50] and 40 [.20]), power dropped to approximately .79. Alexander and DeShon also found that power levels did not decrease as much when sample sizes were equal across moderator-based groups.

In many, and perhaps the majority, of moderator analyses in the social sciences, sample sizes are unequal across moderator-based subgroups. In addition, often the subgroup with the larger sample size is paired with the larger error variance. Test validation research in a variety of organizational settings is a good example of this situation (Hunter, Schmidt, & Hunter, 1979). Typically, the majority subgroup (e.g., Whites, men) is more numerous than the minority subgroup (e.g., African Americans, women), and the majority subgroup has a validity coefficient that is smaller than that of the minority subgroup.

In short, what are the effects of violating the homogeneity of error variance assumption on Type II error rates and, perhaps more relevant for MMR users, on statistical power? When the subgroup with the larger sample size presents the larger error variance (i.e., smaller correlation coefficient), Type II error rates are increased, power is decreased, and MMR users are likely to *incorrectly* dismiss a moderating effect hypothesis.

$$\rho_{xy} = \frac{\sigma_{xy}}{\sigma_x \sigma_y}$$

VIOLATION OF THE ASSUMPTION IN PUBLISHED RESEARCH

As noted earlier, Aguinis et al. (1999) identified 87 articles that used MMR from 1987 to 1999 in *Academy of Management Journal, Journal of Applied Psychology*, and *Personnel Psychology*, and then assessed the frequency of violation of the assumption in these published MMR analyses. Disappointingly, only eight articles (i.e., 9.2%) provided the descriptive statistics necessary to compute error variance within each group and assess compliance with the assumption. On the other hand, many of the 87 articles reported more than one MMR analysis. Therefore, Aguinis et al. were able to check compliance with the homogeneity assumption for a total of 117 MMR tests.

A major finding of this review was that the assumption was violated quite frequently. The next section of this chapter will describe two criteria used to determine whether the assumption has been violated (i.e., a formal test and a rule of thumb). Only tallying results for which both criteria agreed on whether there was a violation, results of Aguinis et al.'s (1999) review showed that the assumption·was violated in 46 instances (i.e., 39%). If results are counted for which at least one criterion

$$\sigma_{xy} = \frac{\Sigma(x_i - \mu_x)(y_i - \mu_y)}{N}$$

indicated that the assumption was violated, this number increases to 68 (i.e., 58%). Thus, violation of the assumption occurred in approximately 40% to 60% of the MMR tests. Once again, the articles reviewed by Aguinis et al. were published in three of the most influential journals in applied psychology and management. *Academy of Management Journal*, *Journal of Applied Psychology*, and *Personnel Psychology* are known for enforcing rigorous methodological standards. Thus, if approximately 50% of MMR tests reported in these journals violate the homogeneity assumption, this number is likely to be at least as high for the MMR tests reported in other journals in the social sciences.

Given that the simulation work described in the previous section has demonstrated that violating the assumption can affect MMR conclusions, it may be warranted to revisit some substantive research conclusions based on MMR under heterogeneity of variance. For example, one of the articles reviewed by Aguinis et al. (1999), in which the assumption was violated, investigated whether the relationship between the criterion "age at which one learned to swim" and the predictor "number of summers worked" was moderated by ethnicity. In this article, results of MMR's F test indicated that ethnicity was in fact a moderator. However, a perusal of the descriptive statistics reported in this article show that the group with the largest sample size (i.e., Whites) was paired with the smaller error variance. As described earlier, Dretzke et al. (1982) demonstrated that this inverse pairing of n with error variance typically leads to inflated Type I error rates. MMR should not have been used in the first place because the assumption was violated. Aguinis et al. computed the A and J statistics that are appropriate for heterogeneity situations (the A and J statistics are described later in this chapter). Results indicated that, contrary to the results based on MMR reported in this article, ethnicity was not a moderator ($A = 2.59$, $p > .05$; $J = 2.62$, $p > .05$). Based on Dretzke et al.'s results, it is possible that the statistically significant moderating effect reported based on MMR is a product of a Type I error. That is, the finding that ethnicity is a moderator is likely to be erroneous.

There are numerous other illustrations of how substantive conclusions published in some of the best journals in the social sciences may need to be revisited because of a violation of the homogeneity of error variance assumption. Aguinis et al. (1999) discussed another study in which researchers investigated various factors hypothesized to affect the career path of women in nontraditional occupations. This study used MMR to investigate the possible moderating effect of gender in a sample of carpenter apprentices. One of the MMR tests included coworker acceptance as a predictor and union satisfaction as a criterion. Coworker acceptance was operationalized as the extent to which carpenter apprentices are accepted by fellow apprentices, feelings of fitting in, and extent of teasing and harassment experienced (reverse coded), whereas union satisfaction was operationalized as the extent to which carpenter apprentices were satisfied with what the union provided in contract negotiation, job security, service to members, improved wages, and so forth.

Aguinis et al. (1999) independently computed error variances in the female and male groups and found that the homogeneity assumption had been violated. Results of the MMR analysis reported in the original article revealed that the F test was not statistically significant, and the substantive conclusion was that there was no moderating effect of gender on the coworker acceptance–union satisfaction relationships. Once again, MMR should not have been used because the assumption was violated. Moreover, the descriptive statistics used to investigate this relationship were such that the group with the smaller sample size (i.e., women) was paired with the smaller error variance (i.e., greater predictor–criterion correlation coefficient). As noted earlier, violating the assumption in a direct pattern of sample size–error variance situation leads to a reduction of statistical power (Alexander & DeShon, 1994). Aguinis et al. (1999) used the more appropriate A and J statistics and found that, contrary to the conclusion derived by using the inappropriate F test, there was a moderating effect of gender ($A = 4.49$, $p < .05$; $J = 4.20$, $p < .05$). More precisely, the coworker acceptance–union satisfaction slope was steeper for women than for men. That is, the same degree of coworker acceptance led to greater union satisfaction for women than for men, showing a moderating effect for gender. In this article, it is likely that violating the homogeneity assumption led researchers to incorrectly conclude that gender was not a moderator.

Changes in substantive conclusions due to the violation of the homogeneity assumption have implications for theory development. In some cases, a change in substantive conclusions can also affect the implementation of social science interventions (i.e., action research; Aguinis, 1993). Take, for instance, the case of the article just described, addressing factors hypothesized to promote women's success in nontraditional occupations. The moderating effect of gender can be explained by the fact that women in nontraditional occupations likely have different expectations than men do (Aguinis & Adams, 1998). Many women may anticipate and expect being harassed, teased, and ridiculed, whereas men do not. Thus, they expect not to be easily and rapidly accepted. Consequently, a similarly moderate level of perceived acceptance for women and men may just meet men's expectations, whereas it may far exceed women's expectations. In turn, women's exceeded expectations regarding coworker acceptance may create positive affect that spills over to greater union satisfaction scores as compared to men's reported union satisfaction. Future research could explore this post hoc explanation. But, the important conclusion is that an incorrect assessment about the presence of the moderating effect of gender is likely to lead to inefficient, and often counterproductive, social science interventions.

In sum, the homogeneity of error variance assumption is violated very frequently. In fact, a selective review of three highly prestigious social science journals has shown that the assumption has been violated in approximately 50% of articles. Given the widespread violation of the homogeneity assumption, it is likely that substantive conclusions of numerous published studies may change if, instead of MMR, more appropriate statistics had been used to examine the presence of moderating effects.

HOW TO CHECK WHETHER THE HOMOGENEITY ASSUMPTION IS VIOLATED

The previous section pointed out that numerous MMR analyses have been conducted in the presence of heterogeneity of error variance. However, it is also the case that in some research situations the assumption is not violated (e.g., Oswald et al., 2000). How can MMR users check whether the assumption is violated in their specific research situation? As we have seen in Figures 4.1, 4.2, and 4.3, a visual examination of potential differences between the width of scatter clouds can give us some information. But, as is the case with tests of other statistical assumptions, researchers have the choice of using formal statistical tests or heuristic guidelines (Weinzimmer, Mone, & Alwan, 1994).

The first type of assessment includes a formal statistical test. As was noted previously, the homogeneity of error variance assumption in MMR is equivalent to the homogeneity of variance assumption in ANOVA. Several tests have been developed to assess homogeneity of error variance in ANOVA models. These statistical procedures assess whether the null hypothesis of homogeneity of error variances is rejected based on sample information. Many of the tests originally developed for the ANOVA context can be modified to assess compliance with the assumption in the MMR context. Based on error rate comparisons of several of these tests by Games, Winkler, and Probert (1972) and Gartside (1972), DeShon and Alexander (1996) concluded that Bartlett's (1937) M test is one of the best choices available (see Appendix A for the computation of M. S. Bartlett's M). Specifically, Games et al. (1972) and Gartside (1972) found that in simulation conditions applicable to typical research conditions in social science research (e.g., three or fewer subgroups, unequal subgroup sample sizes), Bartlett's M test adhered most closely to nominal Type I error rates and demonstrated the highest statistical power rates. That is, in most cases, the null hypothesis of homogeneity of error variances was rejected when it should have been rejected, and it was not rejected when it should not have been rejected based on the way the population scores had been generated. However, there is one caveat. Bartlett's M is adversely affected by deviations from normality (Games et al., 1972). Consequently, a rejection of the null hypothesis of homogeneity of error variance may be due to deviations from normality and not necessarily to a violation of the homogeneity assumption.

In addition to using statistical tests, an alternative for assessing whether one's data set is complying with the homogeneity of error variance assumption is to use an empirically derived rule of thumb. DeShon and Alexander (1996) conducted a Monte Carlo study regarding the accuracy of MMR to estimate the moderating effect of a categorical variable. They manipulated thousands of parameter values for subgroup sample sizes, various degrees of deviation from homogeneity of error variance, departures from Y normality within each subgroup, and within-subgroup correlation between predictor and criterion scores. Based on the results of this large-scale simulation study, the general conclusion was that the F statistic used in MMR begins to be adversely affected when the error variance in one subgroup is approximately 1.5 times larger than the error variance in another subgroup. Thus, this empirically derived 1.5 "rule of thumb" can also be used for determining whether heterogeneity of error variance is likely to affect MMR-based conclusions.

Thus, there are two less-than-perfect methods for assessing whether we are complying with the homogeneity of error variance. First, Bartlett's M statistic is an extrapolation from the ANOVA context and performs well unless the data deviate from normality. Second, DeShon and Alexander's 1.5 heuristic was derived from a very large simulation study but nevertheless may not apply to every possible research situation. So, practically speaking, how do we assess whether we are violating the assumption?

The best approach is to adopt a combination of formal procedures (e.g., Bartlett's test) and heuristics (i.e., the 1.5 DeShon and Alexander rule of thumb). The recommendation is that both procedures be used and that convergence be sought (Aguinis & Pierce, 1998a). Of course, if conclusions based on both the formal and heuristic methods converge, then MMR users would be more confident about their outcome (i.e., whether the assumption is violated). Fortunately, Aguinis et al. (1999) found that the two criteria agree regarding the compliance with the as-

$$F = \frac{MSR}{MSE} \rightarrow \text{regression}$$
$\rightarrow \text{error}$

sumption in over 80% of the cases. In the approximately 20% of cases for which the criteria provided conflicting results, if one of the criteria showed violation of the assumption the other one fell just short of also suggesting a violation. In short, in general the two criteria yield a similar conclusion regarding whether the homogeneity of error variance assumption has been violated. The next section of the chapter provides recommendations on what to do if the assumption is violated.

WHAT TO DO WHEN THE HOMOGENEITY OF ERROR VARIANCE ASSUMPTION IS VIOLATED

MMR's F test likely provides erroneous results regarding the presence of moderator variables when the homogeneity of error variance assumption is violated (Aguinis & Pierce, 1998a). Assuming that an MMR user implements the assessment techniques described earlier (i.e., Bartlett's test and DeShon and Alexander's rule of thumb) and concludes that error variances are heterogeneous, a practical concern is what to do next. We still need to test the substantive hypothesis that there is a moderator effect. However, a violation of the homogeneity assumption means that we cannot trust MMR results. What do we do?

One nonparametric and three parametric alternatives have been investigated using Monte Carlo simulations for evaluating subgroup regression slope differences for situations involving error variance heterogeneity. The parametric alternatives correct for degrees of freedom associated with the more common MMR F test. The nonparametric alternative does not require that the homogeneity assumption be met. These four alternatives are the following: (1) the Welch–Aspin F^* approximation (Dretzke et al., 1982), (2) James's (1951) second-order approximation (J) (DeShon & Alexander, 1996; see Appendix B for equations), (3) Alexander's normalized-t approximation (A) (Alexander & Govern, 1994; see Appendix C for equations), and (4) the nonparametric chi-square test U (Marascuilo, 1966).

Aguinis and Pierce (1998a) reviewed Monte Carlo studies that investigated the accuracy of each of these alternative tests. To summarize their review, the A and J statistics are the best alternatives to the more traditional MMR F test in conditions of heterogeneous error variances for evaluating a moderator variable hypothesis. Although the A and J statistics are more susceptible to deviations from normality (Wilcox, 1997a) as compared to the U statistic, the U has little utility. This test does not perform well in situations with small or moderately large sample sizes, and it has stringent requirement that the intercepts across the (maximum of 2) subgroups be equal (DeShon & Alexander, 1996;

Wilcox, 1988). The F^* has no advantage over the A or J statistics and also does not perform well when scores are not normally distributed. The A statistic is simpler to compute and is slightly more robust to deviations from normality. However, one important advantage of J is its greater statistical power in small-sample situations, which are fairly typical in social science research. Finally, the statistical power of the nonparametric U statistic is not as satisfactory as that of the A and J parametric options. The next section describes a computer program that performs all calculations needed to solve the equations shown in Appendixes A, B, and C to assess whether the assumption has been violated and to produce the A and J statistics.

In addition to the preceding strategies, researchers could implement a weighted least-squares (WLS) regression in lieu of the typical ordinary least-squares regression analysis (Overton, 2001). This correction strategy yields results that are similar to the A and J statistic but has the limitation that it is only applicable to situations involving a binary moderator variable. On the other hand, because the overwhelming majority of published studies in the social sciences using MMR with categorical variables involve tests of binary moderators, this does not limit the practical usefulness of the WLS regression approach. Furthermore, one advantage of implementing WLS regression over the computation of A and J is that WLS allows for the computation of confidence intervals around the regression coefficients. A second advantage is that it maintains MMR within the familiar multiple regression framework (Overton, 2001). This allows for follow-up analyses such as a comparison of predicted Y scores at various levels of X across the two moderator-based subgroups. These follow-up analyses allow for fuller descriptions of the form of the interaction effect.

WLS regression operates by restoring homogeneity of error variance when it does not exist. Overton (2001) provided a detailed mathematical description of WLS regression. From a practical standpoint, the following steps are needed to implement this analysis using computer programs (Overton, 2001):

1. Conduct a bivariate OLS regression using X as the predictor and Y as the criterion for each of the groups.

2. Compute each group's weight by dividing $df - 2$ (df [degrees of freedom] are shown on the output summary table) by the residual sum of squares (labeled "Residual" and also listed on the output summary table). In symbols, $W_i = (df_i - 2)/SS_{res_i}$ is the equation.

3. Use the compute statement to create a new "weight" variable set equal to W_1 for all observations in Group 1 and W_2 for all observations in Group 2 based on the results of Step 2.

4. Conduct the regression analysis exactly as described in Chapter 2, but (if using SPSS specifically) select REGWGT in the REGRESSION command and specify the "weight" variable name.

The resulting WLS regression analysis yields identical regression coefficients for the intercept and for X, Z, and the $X \cdot Z$ product term, but the coefficients' standard errors differ from the typical OLS regression results. The discrepancy between the WLS and OLS regression results increases as the error variances across groups become more divergent (Overton, 2001).

Finally, it may seem that a possible way to go around the homogeneity of error variance assumption is to regress the criterion (e.g., GPA) on the quantitative predictor (e.g., test scores) separately for each of the groups and then compare the resulting regression coefficients (e.g., using the multiple-group procedure in structural equations modeling software packages). This strategy does not really avoid either the problem that the assumption may have been violated or the subsequent impact on Type I and Type II error rates described earlier. More than half a century ago, Gulliksen and Wilks (1950) suggested that a test for differences in standard errors of the estimate (i.e., square root of within-subgroup error variances) should be conducted before testing for differences in slopes across subgroups. Similarly, Einhorn and Bass (1971) also suggested that researchers test for differences in standard errors of estimate across groups. Moreover, Gulliksen and Wilks recommended that tests for inequality of slopes not be conducted in situations involving heterogeneity of standard errors of the estimate across subgroups. In short, investigators should be concerned about satisfying the homogeneity of error variance assumption even if the moderator test involves a between-group comparison as opposed to the traditional MMR test.

ALTMMR: COMPUTER PROGRAM TO CHECK ASSUMPTION COMPLIANCE AND COMPUTE ALTERNATIVE STATISTICS IF NEEDED

A friend of mine presented me with the following, in joke form:

Three researchers walk into a data bar and see a stack of computer code and printouts on a table. There's a quantitative psychologist, a clinical psychologist, and an industrial/organizational psychologist. There is an important statistical assumption needed before the substantive analysis can be conducted. Which psychologists conduct the substantive analysis, which don't, and why?

Answer: The quantitative psychologist knows the assumption well and also knows that it has been violated; therefore, she decides not to conduct the substantive analysis. The clinical psychologist does not know the assumption, so he conducts the statistical test without even checking if the assumption has been violated. The industrial/organizational psychologist knows the assumption well and knows that the assumption has been violated, but conducts the substantive analysis anyway. (Kurt Kraiger, personal communication, January 2002)

The specialty areas of the three characters could probably be replaced with numerous other research areas, and the point is not to pick on any specific field. It just suggests that the mere knowledge of the existence of the homogeneity of error variance assumption, and even of the detrimental effect of violating it, does not suffice for researchers to do something about it. As described by Aguinis et al. (1999), only one of the 87 articles included in the review reported any information regarding whether authors checked for compliance with the homogeneity of error variance assumption. This may be because of a lack of knowledge about it or failure to take the assumption seriously. For example, Weinzimmer et al. (1994) concluded that researchers usually neglect basic regression assumptions and requirements although these are taught in every graduate-level statistics course. Another possible reason for the disregard of the assumption is that there may be a perception that checking for compliance may be too complicated and time-consuming. Aguinis et al. (1999) developed a computer program that makes checking for compliance with the assumption and computing the A and J statistics relatively easy. The program is available at *http:// carbon.cudenver.edu/~haguinis/mmr/*.

Program Description

The program ALTMMR developed by Aguinis et al. (1999) is available in two versions: (1) a browser applet version and (2) a stand-alone version. The program requires minimal input (i.e., sample descriptive statistics rather than raw data) and generates the following output:

1. Ratio of the largest to the smallest error variance, to be compared with the DeShon and Alexander 1.5 rule of thumb to assess homogeneity
2. Bartlett's M test to assess homogeneity
3. James's J statistic to test the null hypothesis that slopes are equal across moderator-based subgroups

4. Alexander's *A*, also to test the null hypothesis that slopes are equal across moderator-based subgroups

In addition, the program can be used as an instructional tool because it includes extensive information regarding MMR, the homogeneity of error variance assumption, and how to interpret the program's output. This information is available via hyperlinks. Thus, investigators could use ALTMMR to deliver live demonstrations of how specific research situations violate the error variance assumption. Moreover, instructors could conduct live demonstrations of how, when the assumption is violated, results from MMR's *F* test are inconsistent with the more appropriate *A* and *J* statistics. Instructors teaching MMR (e.g., Kowalski, 1995) can demonstrate how substantive conclusions published in leading journals may be the result of not complying with the homogeneity assumption, and this demonstration can be a powerful teaching tool that leaves a long-lasting impression.

Input

ALTMMR prompts the user for the necessary information in two steps. First, the user enters the number of moderator-based subgroups included in the study (k) and selects the nominal Type I error rate (i.e., α). The a priori alpha level is only necessary for James's *J* because the program calculates the precise *p* values for the *M* and *A* statistics. Second, the user enters information regarding sample size, correlations, and standard deviations for each moderator-based subgroup.

Assessment of Variance Heterogeneity

ALTMMR assesses whether the homogeneity of error variance assumption is violated by computing Bartlett's *M* statistic and the error variance ratio to be compared to DeShon and Alexander's (1996) 1.5 rule of thumb. The user must provide (1) the number of subgroups (k), (2) the standard deviation of the criterion *Y* for each subgroup (i.e., s_{XY_i}), (3) the correlation between the criterion *Y* and the predictor *X* for each subgroup (i.e., r_{XY_i}), and (4) the sample size for each subgroup. Each of these statistics is easily obtainable using any of the widely available statistics software packages (e.g., SAS, SPSS). For example, Chapter 3 showed how to implement the Frequencies procedure using SPSS. Based on this information, the program computes Bartlett's *M* and the precise associated *p* value of rejecting the null hypothesis that the variances are equal.

To compute the error variance ratio, ALTMMR uses $s_{Y_i}^2$ and r_{XY_i} as estimates for $\sigma_{Y_i}^2$ and ρ_{XY_i}, respectively. Then, each subgroup error variance is computed using Equation 4.1 (note that Equation 4.1, based on parameters, and Equation 4.2, based on sample statistics, yield similar results as sample size increases). When more than two subgroups are evaluated, the program selects the largest error variance ratio from all possible pairwise combinations of error variances. The resulting error variance ratio is compared to the 1.5 rule of thumb.

Computation of Alternatives to MMR

The J statistic, its associated adjusted critical value, and the A statistic are calculated using the equations shown in Appendixes B and C, respectively. The J statistic was originally developed for testing the equality of k independent means in the presence of heterogeneity of variance. Subsequently, J was adapted by DeShon and Alexander (1994) to test for the equality of regression slopes. The computation of J entails calculating a U statistic and then correcting the degrees of freedom used to reference U to the chi-square distribution. The A statistic approximates a chi-square distribution with $k - 1$ degrees of freedom (k is the number of moderator-based subgroups), and is based on a normalizing transformation of the t statistic (DeShon & Alexander, 1994). Thus, the value of J is not referenced to the chi-square distribution directly (as is the A statistic). The adjusted critical value for J is calculated from an initial critical value based on the actual degrees of freedom for the sample.

Output

Figure 4.4 shows the output screen produced by ALTMMR for hypothetical data including two moderator-based subgroups. As shown in this illustrative output screen, ALTMMR first outputs the user-input values to enable verification of the accuracy of the data entered. Second, the program provides information regarding compliance with the homogeneity of error variance assumption as follows: (1) the (largest) ratio of error variance and (2) the value of Bartlett's M statistic and its associated p value regarding the null hypothesis of homogeneity of variance. In this hypothetical example, both indicators suggest that the error variances are heterogeneous across subgroups.

In addition to providing the homogeneity-related results, the program includes text to aid in the interpretation of the results. Note that the violation of the assumption is a matter of degree rather than a binary outcome. Therefore, if (1) the error variance ratio falls just short of

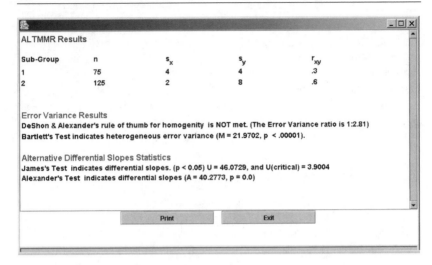

FIGURE 4.4. Sample output screen for illustrative use of ALTMMR program.

the 1:1.5 critical value, (2) the p value associated with the M statistic falls just short of the preestablished Type I error rate (e.g., .05), or (3) there is a discrepancy between the two results, Aguinis et al. (1999) provided the following advice: In addition to reporting MMR's F test, researchers should report the A and J values. If all three statistics (i.e., F, A, and J) lead to the same conclusion regarding the presence of a moderating effect, researchers can be confident about their results. Alternatively, if results of the three tests do not converge, researchers should report results regarding compliance with the assumption and acknowledge that results must be replicated before definitive conclusions are made. Obviously, this second scenario introduces uncertainty and does not allow MMR users to make a decisive conclusion regarding their moderating effect hypothesis. However, the same scenario takes place when the statistical power of MMR's F test is low. More precisely, Aguinis and Stone-Romero (1997) suggested that when an MMR analysis is conducted at low levels of statistical power, researchers should acknowledge that results must be replicated before concluding that a null hypothesis of no moderating effect is tenable.

In addition to the homogeneity-related results, Figure 4.4 shows that ALTMMR displays results regarding the J and A statistics (i.e., alternatives to MMR in case of heterogeneity). Conditional statements also provide information regarding the interpretation of these values. That is, the user is provided with the adjusted critical value for J, the precise

p value for A, and whether these values indicate the presence of a moderating effect.

In sum, the output screen provides information regarding (1) the assessment of compliance with the homogeneity of error variance assumption, (2) alternative tests to MMR, and (3) information regarding these results that aid researchers with interpretation.

CONCLUSIONS

■ This chapter described the homogeneity of error variance assumption, and it distinguished it from the perhaps more familiar homoscedasticity assumption (i.e., constant distribution of scores throughout the regression line). The homogeneity of error variance assumption can be violated even if the usual homoscedasticity assumption is met for all the scores combined or for each of the moderator-based subgroups. Violating the homogeneity of error variance assumption can make MMR's F test results erratic. Type I error rates can be inflated or overly conservative. On the other hand, statistical power can reach unacceptably low levels. Therefore, when the assumption is violated, substantive research conclusions are likely to be erroneous. In short, MMR's F test can be misleading and, therefore, should not be used in the absence of homogeneity of error variance.

■ Results of a selective literature review showed that the assumption is violated in approximately 50% of published studies that use MMR to assess the moderating effects of categorical variables. This is an unfortunate finding because violating the assumption may lead to erroneous substantive conclusions. In turn, these erroneous substantive conclusions are likely to lead to erroneous theory and social science interventions that may create results opposite to what is intended.

■ This chapter also included the description of alternatives that can be used when the assumption is violated, including the A and J statistics and a WLS approach. The chapter described and illustrated the use of the computer program ALTMMR, available on the Web, that allows for the assessment of whether the assumption has been violated. The program also allows for the computation of the alternative statistics A and J that can be used in lieu of MMR's F test to ascertain whether there is a moderating effect. This program can be used in the classroom to illustrate the effects of violating the assumption on substantive conclusions about moderator variables.

Given the adverse effects of violating the homogeneity assumption and the ease of use of the ALTMMR program, it is hardly justified for MMR users not to check for compliance with the homogeneity assumption before conducting a moderator test.

■ The next chapter will address a major potential problem faced by MMR users. Specifically, the next chapter discusses the numerous factors that adversely affect the statistical power of MMR.

5

MMR's Low-Power Problem

Statistical power is good; absolute concern for statistical power corrupts absolutely.
—SECHREST (2000)

As was noted in Chapter 1, conducting an MMR analysis is much like walking in a minefield. There are numerous hidden obstacles that researchers must dodge to reach the goal of completing the analysis successfully. For the most part, these hidden obstacles cause MMR to suffer from low statistical power. In practical terms, this means that a hypothesis about the presence of a moderator variable may be correct, but a researcher concludes *incorrectly* that there is no moderating effect. This chapter frames the issue of statistical power within the broader topic of statistical inference. It also describes briefly the controversy over null hypothesis significance testing (NHST). Then it provides a detailed description of each of the methodological and statistical artifacts that affect the power of MMR. This chapter makes painfully evident that the conditions that are adverse to adequate statistical power are very common in most social science research settings. Finally, this chapter reports results of a review of the typically observed magnitude of moderating effect sizes and the relationship between effect sizes and power in published articles. Chapter 6 provides potential solutions to the low-power problem.

STATISTICAL INFERENCES AND POWER

Statistical power is the probability that, given that there is a moderating effect in the population, this effect will be detected in the sample in

65

hand. As described in Chapter 2, MMR tests the null hypothesis H_0: $\beta_3 = 0$, or that the coefficient for the product term is zero in the population, based on the following equation for binary moderators:

$$Y = a + b_1 X + b_2 Z + b_3 X \cdot Z + e \qquad (5.1)$$

In the unlikely situation when a researcher has access to all the population scores, then the MMR analysis actually involves β_3 (i.e., the population regression coefficient) and not just the sample-based b_3, which is typically used as an estimate of β_3. In such a scenario, inferential statistics are not needed.

As described in Chapter 2, we need to test the likelihood that the null hypothesis of no moderating effect is true by using a t test based on b_3 or an F test based on ΔR^2 (i.e., the proportion of variance in Y explained by adding the product term to the equation). When we use inferential procedures, we make conclusions about populations based on samples. Based on the MMR analysis using the sample, we can conclude that the null hypothesis is false or that we do not have sufficient evidence to conclude that it is false. In the latter case, we conclude that the null hypothesis is tenable and that it is likely there is no population moderating effect.

Figure 5.1 shows the well-known four possible conclusions researchers can make based on the sample vis-à-vis the actual (unknown) state of affairs in the population. One can conclude correctly that H_0 is false, and one can also conclude correctly that H_0 is tenable. Of course, these are desirable correct decision outcomes. However, one can conclude *incorrectly* that H_0 is false, making what is called a Type I error (α), and one can conclude *incorrectly* that H_0 is tenable, making what is

Conclusion Based on the Sample

		H_0 is false	H_0 is tenable
Actual State of Affairs	H_0 is false	Correct conclusion	Type II error (β)
	H_0 is true	Type I error (α)	Correct conclusion

FIGURE 5.1. Decision outcomes based on the sample vis-à-vis the actual (unknown) state of affairs in the population.

called a Type II error (β). Statistical power equals $1 - \beta$. Thus, the lower the power of the MMR test, the larger β and the possibility that we incorrectly conclude there is no moderator variable.

CONTROVERSY OVER NULL HYPOTHESIS SIGNIFICANCE TESTING

Null hypothesis significance testing (NHST) has been, and still is, a topic of heated debate in the social science community (e.g., Chow, 1996; Cohen, 1994; Cortina & Dunlap, 1997; Cortina & Folger, 1998; Markus, 2001; Murphy, 2002, Nickerson, 2000; Task Force on Statistical Inference, 2000; Tryon, 2001; Wilkinson & the Task Force on Statistical Inference, 1999). Some argue that significance testing is useful (e.g., Wainer, 1999), whereas others believe that it is misleading and should be discontinued (e.g., Schmidt, 1996). Therefore it is important to clarify the purpose and meaning of NHST as well as a likely reason for the controversy.

Purpose of NHST

The purpose of NHST is to determine whether a finding of a moderating effect can be explained by chance alone (i.e., sample fluctuations) (Cascio & Aguinis, 2001, 2003). As noted earlier, significance testing is used only when we use samples to make inferences regarding populations. Therefore, NHST is not needed when we do not wish to make inferences from samples to populations. However, the more typical situation in social science research is that we have a sample of scores and would like to draw conclusions about the population of scores. In this case, we need to infer whether, for example, $b_3 = .56$ can be explained by chance alone (i.e., sample fluctuations) or is a robust finding. In other words, if we replicated the same study in a separate sample, is it possible that results would show that $b_3 = 0$? In this situation, in which we make inferences from samples to populations, the purpose of significance testing is to provide information regarding whether the value found for b_3 can be explained by likely differences in the population or by chance alone.

Meaning of NHST

Rejecting the null hypothesis means that we believe the data are different from what they would be if there were no moderating effect (Cascio & Aguinis, 2001, 2003). However, rejecting the null hypothesis does not inform us about the magnitude of the presumed moderating effect, or even it if is in the right direction (i.e., whether the slope of Y on X is steeper for one or the other level of the moderator Z). All we can infer is that, based on sample information, there is a moderating effect. Moreover, NHST does not allow us to make a statement regarding how large the effect is and what are the underlying mechanisms causing it. Without knowing the moderating (effect size) it is difficult to establish whether the effect is "practically significant." It is likely that all moderating relationships are different from zero, at least at the "nth" decimal place. But is a statistically significant moderating effect based on $b_3 = .0001$ a meaningful finding for science and practice? NHST does not answer this question, but only whether we believe the effect is not zero. Chapter 9 discusses several ways of assessing whether, in addition to being statistically significant, the moderating effect found is also practically significant.

Not rejecting the null hypothesis means that, even if $b_3 \neq 0$ (or $\Delta R^2 > 0$) is observed in the sample, we cannot rule out the possibility that the observed sample effect is due to chance alone and, in fact, the moderating effect is zero in the population. In addition, not rejecting the null hypothesis does not mean that the moderating effect is zero. It just means that, based on the sample information we have, there is not sufficient empirical evidence to conclude that an effect exists. In fact, it is possible that there is a moderating effect in the population but this effect may be undetected in the sample, mainly due to inadequate statistical power (Aguinis, 1995; Aguinis & Stone-Romero, 1997). As shown in Figure 5.1, this is a Type II error.

Controversy over NHST: A Human Factors Problem

Human factors is a field concerned with the performance of people in task-oriented environments interacting with equipment, other people, or both. It seems that the controversy surrounding NHST is due mainly to how significance testing is used. In other words, many researchers have noted that significance testing is abused and misused (e.g., Cohen, 1994; Schmidt, 1996) and, therefore, there seems to be a "human factors problem" (Tryon, 2001, p. 371). Significance testing allows researchers to infer whether the null hypothesis of no moderating effect is likely to be false. On the other hand, significance testing is used incorrectly when (1) conclusions are made regarding the magnitude of the moderating effect (e.g., a statistically significant result at the .01 level is interpreted as a larger moderating effect than a result at the .05 level), and/or (2) failure to reject the null hypothesis is interpreted as evidence that the moderating effect does not exist (i.e., because not detecting a moderating effect in the sample may be due to insufficient statistical power). Next is a description of the factors affecting the power of inferential tests in general and MMR in particular.

FACTORS AFFECTING THE POWER OF ALL INFERENTIAL TESTS

Cohen (1988) suggested that social science researchers should strive to achieve a statistical power level of at least .80. That is, if there is a population moderating effect, researchers should be able to conduct an MMR analysis that has an 80% chance of concluding that there is a moderator. However, there are many factors that prevent researchers from achieving this desired level of statistical power. The following three factors are known to affect the power of all inferential tests, including MMR:

• A decrease in the size of the sample produces a decrease in power. Unfortunately, as will be described later in this chapter, the typical sample size found in MMR tests conducted in social science research is usually inadequate to produce sufficient levels of statistical power (i.e., .80 or greater).

• A decrease in the preset Type I error rate produces a decrease in power. When we conduct our MMR analyses, we choose an α that is conventionally .05. However, given other things equal, the larger the size of α, the greater the resulting statistical power.

• A decrease in the size of the moderating effect in the population produces a decrease in power. On the other hand, a large population effect is likely to result in a large sample-based effect, which in turn results in a greater likelihood that H_0 will be rejected. In short, it is easier to detect the moderator in the sample when there is a large moderating effect in the population. But, as will be described in this chapter, many design, measurement, and analysis artifacts decrease the observed effects as compared to their population counterparts.

The preceding three factors affect the statistical power of all inferential tests. Thus, MMR users need to be aware of their effects. In addition, there are several factors that have been identified over the past few years that also have an important impact on power. Each of these factors, many of them specific to the context of MMR, affects power because it has a direct or indirect impact on at least one of the preceding factors known to affect the power of all inferential tests. These factors are discussed next.

FACTORS AFFECTING THE POWER OF MMR

Aguinis (1995) reviewed factors that affect the power of MMR. Since this review, simulation research has uncovered additional issues that should also be considered. Expanding on this review, the factors affecting the power of MMR can be grouped into the following five categories:

- Variable distributions (predictor variable variance reduction, transformations of skewed criterion scores).
- Operationalizations of criterion and predictor variable (measurement error, scale coarseness, polichotomization of truly continuous variables).
- Sample size (total sample size, sample size across moderator-based subgroups).
- Characteristics of the predictor variables (correlation between the predictor X and the moderator Z, correlation between the predictor X and the criterion Y).
- Interactive effects (i.e., the factors listed earlier have not only additive but also interactive effects on power).

As noted earlier, each of these factors affects power because it has a direct or indirect impact on at least one of the general issues known to affect the power of all inferential tests. For example, predictor variable reduction and measurement error decrease the estimated effect size. Each of these factors is described next.

Variable Distributions

There are two aspects of variable distributions that affect the statistical power of MMR: (1) predictor variable variance reduction and (2) transformations of skewed criterion scores (e.g., logarithm transformation).

Predictor Variable Variance Reduction

In many situations in social science research, the variance of scores in the sample under study is smaller than the variance of scores in the population (Linn, 1968). This discrepancy between the sample and population variances is a type of nonrandom sampling because not all population scores have the same probability of being included in the sample (Alexander, Barrett, Alliger, & Carson, 1986).

There are many research areas in which the population and sample variances are systematically different. Test validation research serves as a good illustration. In concurrent validation designs, data are collected regarding a newly developed test and a criterion (e.g., supervisory ratings of performance) to assess the extent to which the new test predicts criterion scores. However, the criterion scores are likely to be truncated because when the employees were selected to be part of the organization, decisions regarding their selection were based on their standing on various assessment instruments (e.g., measure of general mental abilities, selection interview). In other words, only those who obtained a score that exceeded a specific cutoff point were selected, which led to a reduction of variance in the abilities, personality, and other characteristics of the individuals selected for the job vis-à-vis the individuals in the population of applicants. Therefore, the variance on these variables is likely to be smaller in the sample in hand than in the population.

Returning to the MMR model shown in Equation 5.1, the power of MMR is reduced markedly when the variance of X is smaller in the sample than in the population (Aguinis & Stone-Romero, 1997) (Chapter 6 describes the effects of variance enhancement, as opposed to variance reduction). Results reported by Aguinis and Stone-Romero (1997) demonstrated that by restricting the range of X scores, the variance of $X \cdot Z$ scores was also restricted, and this adversely affected the ability of MMR to detect a population moderating effect. Moreover, although Aguinis and Stone-Romero investigated direct truncation on X, truncation on variables that are correlated with X (i.e., indirect truncation) is also pervasive in social science research (Aguinis & Whitehead, 1997; Hunter, Schmidt, & Le, 2002) and hence also lower the power of MMR.

What type of difference between the population and sample variance on X is sufficient to have a noticeable impact on the power of MMR? Aguinis and Stone-Romero's (1997) Monte Carlo study revealed that a surprisingly mild ratio of sample to population variance can have substantial effects on power. For example, in a situation with $N = 300$ and no truncation, the power to detect what Aguinis and Stone-Romero operationalized as a medium-size moderating effect was .81. However, when the scores were sampled from only the top 80% of the distribution of the population scores, power decreased to an unacceptable .51. Thus, even in the presence of a relatively mild degree of range restriction (i.e., the bottom 20% of the distribution is truncated), power loss poses a serious threat to the validity of MMR-based conclusions.

Transformations of Skewed Criterion Scores

In many social science domains, researchers are faced with severely skewed criterion scores. Research on monetary compensation, particularly executive compensation, serves as a good illustration (Conyon &

Peck, 1998). More often than not, distributions of compensation scores include a few extreme scores who receive very high values, producing positively skewed distributions. In such situations, researchers often choose to transform the skewed scores by using a log transformation that normalizes the distribution (Tabachnick & Fidell, 1989, pp. 83–87; Winer, 1974, pp. 398–401). Although there are many types of transformations that can be used (e.g., square root, reciprocal), "it could be argued that performing log transformations on positively skewed dependent variables has become a convention within applied psychology and management research" (Russell & Dean, 2000, p. 167).

A key issue in implementing transformations to normalize criterion scores is that the resulting relationship between Y and the predictors may not be linear, violating one of the basic OLS multiple regression assumptions (see Chapter 2). In general, what are the effects on power of log-transforming criterion scores and then performing MMR analyses using the transformed data? To answer this question, Russell and Dean (2000) conducted a large-scale simulation comparing MMR results based on (1) a simple percentile bootstrap procedure based on the positively skewed (untransformed) data and (2) a traditional MMR analysis using log-transformed scores. Bootstrapping consists of simulating the process of randomly selecting samples from a population in order to construct a sampling distribution (Diaconis & Efron, 1983; Efron, 1979). Conclusions based on their simulation, which included a population moderating effect, showed that the sample-based estimated population moderating effects were 3.5–6.9 times larger when computed using the bootstrap procedure based on the (skewed) untransformed data as compared to using MMR with the log-transformed data. They also found that the difference in favor of the untransformed scores was larger as the magnitude of the moderating effect increased. As noted earlier, the size of the moderating effect is one of the key factors that affects power. Thus, an underestimation of the moderating effect produces a decrease in statistical power. Therefore, Russell and Dean (2000) concluded that "it is clear that log transformations of positively skewed dependent variables greatly enhance the probability of Type II error" (p. 177) (i.e., not detecting an existing population effect).

Operationalization of Criterion and Predictor Variables

Researchers are faced with many choices in terms of how to operationalize their predictor and criterion variables. These choices likely have an impact on the power of the MMR test and, consequently, on the accuracy of substantive conclusions regarding the presence of a moderator variable. The following three issues regarding operationalization of predictor and criterion variables are discussed next: (1) measurement error, (2) scale coarseness, and (3) polichotomization of truly continuous variables.

Measurement Error

Researchers searching for the reasons for the low power of MMR have identified measurement error as one of the culprits (e.g., Busemeyer & Jones, 1983; Dunlap & Kemery, 1988; Evans, 1985). In fact, some have argued that measurement error is possibly the most influential factor affecting power (Ganzach, 1998; Kromrey & Foster-Johnson, 1999).

In the social sciences, constructs are virtually never measured with perfect reliability. In the specific context of MMR, the categorical moderator Z is typically measured at high levels of reliability, but this is not the case for the quantitative predictor X. Consequently, the estimated regression coefficients are usually attenuated (Fisicaro & Lautenschlager, 1992). In other words, in situations of less than perfect reliability in the predictor X and moderator Z, the reliability of $X \cdot Z$ scores is adversely affected, and the sample-based b_3 underestimates the population coefficient β_3 (cf. Equation 2.2). Because effect size affects power, an underestimation of β_3 is likely to lead researchers to the incorrect conclusion that the null hypothesis H_0: $\beta_3 = 0$ is tenable.

However, what is the impact of measurement error on the power of the MMR test? Simulations that include situations where both X and Z are quantitative have confirmed Busemeyer and Jones's (1983; see also Bohrnstedt & Marwell, 1978) analytical work that the estimated effect size for the product term is reduced when the reliabilities of the predictors are small (Evans, 1985). In addition, Dunlap and Kemery (1988) examined the effects of reliability in models including quantitative predictors X and Z and their correlation on statistical power, and replicated previous findings by Evans (1985). For example, when the reliabilities for X and Z were .50, for a high correlation between the predictors (e.g., .80), statistical power was .706. However, when the correlation between the predictors was lower (e.g., .20), power decreased to .561. Although Dunlap and Kemery's simulation did not vary sample size (i.e., $N = 30$ for all conditions), research by Paunonen and Jackson (1988) used a similar Monte Carlo design and corroborated Dunlap and Kemery's results for a larger sample size ($N = 100$).

In sum, MMR users should be aware that reliability levels that are perceived as acceptable for testing hypotheses regarding the regression coefficients for X and Z result, in many cases, in substandard reliability levels for the $X \cdot Z$ product term. Furthermore, testing a potential moderating effect with an inadequate level of reliability of measurement is likely to lead to a Type II error.

Scale Coarseness

A second issue regarding the way variables are operationalized that affects statistical power is what has been labeled "scale coarseness" (Bobko & Russell, 1994; Russell & Bobko, 1992; Russell, Pinto, & Bobko, 1991). Researchers who use MMR typically operationalize the quantitative predictor variable X using Likert-type scales including five or seven anchors. Assuming the case of a 5-point scale for X and a binary moderator variable Z such as gender, the product term $X \cdot Z$ has a possible range of $5 \times 2 = 10$ distinct values. The criterion Y is also likely to be measured on a similar scale as X (i.e., Likert-type scale including five anchors). Thus, Y is measured on a "coarse" 5-point scale rather than on the same 10-point scale as the product term. The insufficient number of scale points for the Y scale results in possible information loss and, therefore, may prevent a moderating effect from being detected. The scale coarseness problem is particularly relevant when measurement scales do not contain at least as many response options as exist in the theoretical response domain (Russell & Bobko, 1992).

Russell and Bobko (1992) conducted an experiment in which participants were assigned to one of two conditions. The first condition included a criterion variable operationalized using a 5-point Likert-type scale. The second condition followed a practice used by Arnold (1981) consisting of including a criterion variable operationalized as a graphic line segment on which participants had to place a mark indicating their response. Results of this experiment demonstrated that the estimated size of the moderating effect was larger for the condition in which study participants used the continuous graphic scale as opposed to the coarse 5-point Likert-type scale, and this number of anchors is not at least as large as those in the theoretical response domain.

In sum, the way the criterion variable Y is operationalized can have an impact on whether an existing moderating effect is detected. Even before the data are analyzed, this measurement consideration can have an impact on substantive conclusions about the operation of moderator variables. Some researchers have tried to avoid the scale coarseness problem by using Likert-type scale with as many as 19 anchors (Aguinis, Nesler, Quigley, Lee, & Tedeschi, 1996). Nevertheless, it is not possible to avoid the scale coarseness problem as long as the Y variable is measured on an interval-level scale including the same number of anchors as the scale used to measure the criterion X.

Polichotomization of a Truly Continuous Variable

A common practice for testing hypotheses about moderating effects including at least one quantitative predictor variable consists of dichotomizing the variable into two categories (e.g., "high" and "low") and then conducting an ANOVA using the artificially created groups. Numerous published studies include such an approach to testing moderating effects such that a truly continuous variable (e.g., age) is artificially polichotomized (e.g., Group 1 = under 40, Group 2 = 40 and over). This practice seems to be particularly pervasive in the area of personality psychology (Bissonnette, Ickes, Bernstein, & Knowles, 1990; Chaplin, 1997).

In general, Cohen (1983) demonstrated that artificially dichotomizing (or polichotomizing) truly continuous variables is inappropriate because it reduces the chances of detecting an effect. More recently, Vargha, Rudas, Delaney, and Maxwell (1996) implemented computer simulations demonstrating that the process of dichotomization involves a quantifiable loss of information. Specific to the context of MMR, Stone-Romero and Anderson (1994) and Mason, Tu, and Cauce (1996) demonstrated that artificial dichotomization of a quantitative predictor leads to substantial power loss in tests of moderators. Note that Maxwell and Delaney (1993) ascertained that the simultaneous dichotomization of more than one predictor not only decreases the power of MMR but, in some situations, may also artificially increase Type I error rates (Baumeister, 1990; Bissonnette et al., 1990).

In sum, researchers should not use procedures such as a median split to artificially create subgroups based on a variable measured using a continuous scale. In most cases this is likely to reduce the chances of detecting a moderating effect, and in others it may lead to an increase in Type I error rates. In all cases, however, artificial polichotomization of continuous variables is likely to lead to incorrect substantive conclusions regarding the presence of a moderating effect.

Sample Size

There are two issues regarding sample size that affect the power of any MMR analysis. The first one is the total size of the sample (N). The second one is the relative sample size in each of the moderator-based subgroups (n_i). Following is a description of these factors and their effects on statistical power.

Total Sample Size

Sample size is a factor that is positively related to the statistical power of any inferential test. The size of the sample on which the MMR analysis is performed is one of the most important single factors affecting power. In recent years, several Monte Carlo simulations have explored the effects of sample size on the power of MMR to detect moderators (e.g., Alexander & DeShon, 1994; Stone-Romero & Anderson, 1994). Stone-Romero and Anderson, for example, found that what they defined as a small effect size was typically undetected when sample size was as large as 120, and that unless a sample size of at least 120 was used, even what they defined as medium and large moderating effects were, in general, also undetected.

In sum, users of MMR should be aware that size of the sample in hand is one of the most important determinants of statistical power. Thus, a small sample (< 100) is a virtual guarantee of not being able to detect a moderating effect.

Sample Size across Moderator-Based Subgroups

In most research situations in the social sciences, the moderator-based subgroups are dissimilar in size (e.g., more Whites than Latinos and African Americans; more members of one gender-based group than the other) (Hunter, Schmidt, & Rauschenberger, 1984). In general terms, Hsu (1993) showed that the *effective* total sample size (N') for two independent sample tests of means, correlations, and proportions is the harmonic mean of the two subgroup sample sizes:

$$N' = \frac{2(n_1 n_2)}{n_1 + n_2} \tag{5.2}$$

Consequently, in unequal subgroup sample sizes situations (i.e., $n_1 \neq n_2$), when the size of one of the subgroups is fixed at n_1, the statistical power of an inferential test cannot exceed the power of a test involving two subgroups, each of size $2(n_1)$, regardless of the size of the second subgroup. To use an example, assume a test involves a group of 40 African Americans and 300 whites. One may think that power would be based on a total sample size of 40 + 300 = 340. However, power cannot exceed the power of a test including a total sample size of 160 because $2(n_1) + 2(n_1) = 2(40) + 2(40) = 160$.

What is the actual effect of unequal subgroup sample sizes on the power of MMR? In situations with two subgroups, there is a considerable decrease in power when the size of Subgroup 1 is .10 relative to total sample size, however large or small the latter may be (e.g., 30, 60, 180, 300) (Stone-Romero, Alliger, & Aguinis, 1994). The effect of unequal subgroup proportions on statistical power is significant *above and beyond* the effect of total sample size. A proportion of .30, closer to the optimum value of .50, also reduces the statistical power of MMR, but to a lesser extent (Stone-Romero et al., 1994).

Characteristics of the Predictor Variables

There are two characteristics of the predictor variables that have been described as related to the power of MMR to detect moderating effects. These are the strength of the correlation between (1) the predictor X and the moderator Z, and (2) the predictor X and the criterion Y. Each of these relationships and its effects on power is described next.

Correlation between the Predictor X and the Moderator Z

Some researchers (e.g., Morris et al., 1986; Smith & Sasaki, 1979) have argued that the presence of multicollinearity (i.e., correlation between the predictor X and the moderator Z) leads to an ill-conditioned solution in which the regression coefficients are unstable, error terms are increased, and the power of MMR is consequently decreased. In the MMR model, there may be multicollinearity between the predictor X and the moderator Z. In addition, the product term $X \cdot Z$ is always correlated with its components X and Z. Thus, the fact that multicollinearity is guaranteed in MMR has been proposed as a factor that has a detrimental impact on power (Morris et al., 1986).

Fortunately, the concerns about the detrimental effects of multicollinearity on MMR have proven to be unfounded. Cronbach (1987) provided evidence that in the typical MMR analysis including a predictor, a moderator, and the product term, multicollinearity does not adversely affect power. A high degree of multicollinearity may cause problems regarding the interpretation of the coefficients associated with the predictors, but centering takes care of this potential problem. (Chapter 3 included a detailed discussion of centering.) In short, the concern that a high degree of correlation between the predictor X and the moderator Z has a detrimental impact on the statistical power of MMR is unfounded.

Correlation between the Predictor X and the Criterion Y

In a special issue of *Organizational Research Methods* devoted to interaction effects (see Aguinis, 2002, for an overview of this issue), Rogers (2002) provided an additional explanation for why interaction effects are so difficult to detect. Specifically, Rogers demonstrated that the size of the relationship between the predictor X and the criterion Y places a mathematical constraint on the size of the interaction effect in the population. In other words, a small relationship between X and Y may prevent a large moderating effect of Z from existing in the population, and thus from being detected in the sample. Therefore, interaction effects are not likely to be found unless there are strong first-order effects. In short, "in order to have a strong ordinal moderation [i.e., the lines for the groups do not cross within the possible range of the values of X], there must be a strong effect to be moderated" (Rogers, 2002, p. 212).

The fact that the strength of the X–Y relationship places a ceiling on the effect size of the moderating effect reminds MMR users of the need to consider theory and measurement issues around first-order effects as an indirect way to improve the detection of moderating effects.

Interactive Effects

The various factors that have been described that affect the power of MMR have interactive effects (Aguinis & Stone-Romero, 1997). For example, the combined effects on power of the simultaneous presence of small total sample size, large measurement error, and unequal sample sizes across the moderator-based subgroups are greater than the sum of the individual effects of these factors. Furthermore, the presence of an unfavorable value on just one of these factors (e.g., predictor variable range restriction) is sufficient to lower power substantially, even if the values for some of the other factors are conducive to high power (i.e., large sample size, similar sample sizes across moderator-based subgroups).

It should be obvious by now that the analogy at the beginning of this chapter between an MMR analysis and a minefield is fully justified. In the majority of studies in which MMR is used, researchers face at least some of the conditions described earlier that reduce the effect size and power. Therefore, we should not be surprised if the observed effect sizes for MMR tests in published research are quite small. But how small are they? Perhaps as low as what Cohen (1988) defined as a small effect size (i.e., $f^2 = .02$)? Could observed effect sizes be even smaller than what is traditionally identified as a "small" effect? This question is answered next.

EFFECT SIZES AND POWER IN PUBLISHED RESEARCH

Chapter 2 described a review of the MMR literature conducted by Aguinis et al. (2003). This review included all issues of the *Journal of Applied Psychology*, *Personnel Psychology*, and *Academy of Management Journal* from 1969 to 1998. The review included three journals only, but these publications are among the most influential scholarly journals devoted to empirical research in applied psychology and management (Starbuck & Mezias, 1996). Thus, they serve as exemplars of top-tier social science journals devoted to publishing original empirical work.

The Aguinis et al. (2003) literature review included all studies that reported using MMR to assess the potential moderating effect of a categorical variable. This resulted in 106 articles including a total of 636 MMR analyses. Not all articles included sufficient information to compute power. Thus, the authors of the original articles were contacted and requested to provide any missing information. However, very few were responsive. This was a disappointing finding, especially given that ethical guidelines in the social sciences indicate that authors should keep their original data for at least 5 years (Aguinis & Henle, 2002; Johnson, 2001). Nevertheless, based on the information collected from the original articles and that provided by their authors, the resulting number of MMR analyses for which Aguinis et al. (2003) were able to compute effect size and power was 261. This number represents 41% of all the published analyses, and Aguinis et al. (2003) reported that there were no noticeable differences between the sample and the population of MMR analyses regarding a number of key features. Therefore, the sample of MMR analyses seems to be representative of the population.

Aguinis et al. (2003) computed the moderating effect size for each MMR analysis using f^2 (Chapter 9 and Appendix D include a more detailed discussion of this effect size metric). The effect size f^2 is the ratio of systematic variance accounted for by the moderating effect relative to unexplained variance in the criterion. Aguinis et al. also computed statistical power based on each effect size. However, they noted that the computation of power based on observed effect sizes (i.e., post hoc power analysis) is not really useful because such power value is a direct function of, and does not provide more information than, the obtained p value for the MMR test (Gerard, Smith, & Weerakkody, 1998; Goodman & Berlin, 1994; Hoenig & Heisey, 2001). In general, the logical flaw in computing post hoc power analysis is to assume that whatever effect size was observed, even if it is trivial, is one a researcher would wish to find statistically significant (Aguinis et al., 2003). Thus, the reasons why Aguinis et al. computed power based on observed effect sizes was to understand how power changes when values for effect sizes change, and to understand how power is influenced by effect size given typically observed sample sizes.

Results showed that the overall mean f^2 for the 261 analyses was .009, with a 95% confidence interval (CI) ranging from .006 to .012. However, because the effect size distribution was highly skewed, the median effect size of .002 is a better descriptor of central tendency. The 25th percentile was .0004 and the 75th percentile was .0053. The overall small effect sizes found did not vary substantially across journals (i.e., *Academy of Management Journal, Journal of Applied Psychology, Personnel Psychology*), type of moderator (e.g., ethnicity, gender), or research domain (e.g., personnel selection vs. other areas, work attitudes vs. other areas). Also, the mean corrected (for measurement error in X and Y) f^2 was .017 and the median was .003. Thus, the measurement error correction did not increase effect sizes by much. This was expected, given that measurement error is only one of the many measurement, design, and statistical artifacts that influence effect size and power and that these artifacts have not only main but also interactive effects. Because journals are more likely to publish articles that include large effect sizes and lead to statistically significant results than those that report smaller effects sizes and statistically nonsignificant results (Rosenthal, 1979), the magnitude of f^2 was likely to be even lower if the review included unpublished studies.

Regarding the relationship between effect sizes and power, results showed that small increases in f^2 yielded substantial increases in power. For example, as noted earlier, comparing the observed versus corrected (for measurement error) median effect sizes increased f^2 by just .001. However, this increase in effect size was accompanied by a substantial power increase of .09 (i.e., 44%).

In spite of the finding that effect sizes are quite small, the Aguinis et al. (2003) review revealed two encouraging results. First, none of the 95% CIs around the mean effect size for the various comparisons included the value of zero. Thus, although observed effect sizes are small, they seem to be greater than zero. Second, the correlation between effect sizes and year of publication was positive and statistically significant, that is, $r(261) = .15$, $p < .05$. Power seems to have improved over time, most likely as a result of improvements in research design and the heightened degree of awareness that journal editors and reviewers have about MMR and the artifacts that influence effect sizes in general.

IMPLICATIONS OF SMALL OBSERVED EFFECT SIZES FOR SOCIAL SCIENCE RESEARCH

The conclusion of the Aguinis et al. (2003) quantitative literature review puts into question the meaningfulness of relying on broad-based standards for defining what is a critical effect size. Specifically, Cohen (1988) suggested that $f^2 = .02$ is a small effect size. However, the median effect size identified in the Aguinis et al. (2003) review is .002. Researchers who have used Cohen's definitions of small, medium, and large effect sizes in conducting power analyses (e.g., Mone, Mueller, & Mauland, 1996; Sedlmeier & Gigerenzer, 1989) have not acknowledged explicitly that these values are actually based on observed effect sizes for specific, and often limited, literature domains. For example, the values derived by Cohen (1962) are based on observed effect sizes computed from articles published in just one volume of the *Journal of Abnormal and Social Psychology*. Furthermore, because Cohen's now-conventional definitions of small, medium, and large effect sizes are based on observed values, they have been revised over time as a consequence of subsequent literature reviews of effect sizes in various domains. For example, for correlation coefficients, Cohen defined .20 as small, .40 as medium, and .60 as large in his 1962 *Journal of Abnormal and Social Psychology* review. However, he changed these definitions to .10, .30, and .50, respectively, in his 1988 power analysis book.

The Aguinis et al. (2003) review showed that even if a researcher hypothesizes what can be conventionally considered a "small" moderating effect size (i.e., $f^2 = .02$) and, moreover, plans the research designs accordingly so that power will be .80 to detect an effect of .02 or larger, power may be insufficient to detect a more realistic effect, given that the median effect is only .002. Therefore, as long as such small effects have a meaningful impact for science or practice within a specific context, the implication is that researchers should conduct a power analysis and plan future research design based on smaller (and more realistic) targeted effect sizes.

An additional implication of the Aguinis et al. (2003) review is that it is likely that many moderators have gone undetected. For example, take the personnel selection literature. There is a long-held conclusion that ethnicity and gender are not moderators of the preemployment test scores–job performance relationship (e.g., Schmidt, 2002; Te Nijenhuis & Van der Flier, 1999). However, as an illustration, one of the articles included in the Aguinis et al. review concluded that there was no gender-based differential prediction for the relationship between a test battery (i.e., table reading, technical reading, industrial measurement, and eye–hand coordination) and an unweighted measure of task performance. As a second illustration also from the personnel selection domain, another MMR test included in the review concluded there was no gender-based differential prediction for the relationship between scores on an experience-based interview and performance ratings. In sum, given the numerous design, measurement, and analysis artifacts that lower observed effect sizes yield lower power, it is necessary to revisit the substantive research conclusions that there is no differential prediction by gender for the relationship between (1) a test battery that includes various types of abilities and task performance and (2) a selection interview and performance.

In general, researchers in several social science fields have documented the lack of success in supporting hypotheses about moderating effects. Such research domains include job design (White, 1978), personality (Chaplin & Goldberg, 1984), the issue of whether some patients have more "aptitude" to respond positively to therapy than others do (Dance & Neufeld, 1988), to name a few. Extrapolating from the personnel selection domain to these and other research areas, it seems reasonable to suggest that past MMR null findings should be closely scrutinized to assess whether they may have been due to low power, as opposed to the absence of a moderating effect in the population. But could it be that theories in all these domains are incorrect in positing the operation of moderators? Could it really be the case that literally hundreds of researchers in dozens of disparate domains in applied psychology and associated fields are wrong? Obviously, this possibility cannot be discarded. However, it seems more likely that the pervasive and often unavoidable design, measurement, and statistical artifacts decrease the observed effect sizes substantially vis-à-vis their population counterparts, causing researchers to commit a Type II error (i.e., incorrectly concluding that the null hypothesis of no moderating effect is tenable). In fact, this seems to be a more plausible explanation given the Monte Carlo evidence reviewed earlier in this chapter.

In addition to the implications for theory development, the low power of MMR has implications for practice. For example, assume a particular social science intervention (e.g., smoking cessation program) has a positive effect on a specific group (e.g., Baby Boomers) and a neutral or even negative effect on members of another group (e.g., Generation Xers). Not detecting this moderating effect of group membership may lead to the ineffective and wasteful decision that the program should be implemented for individuals of all age groups. Of course, numerous additional examples can be used to show that the inability to recognize contingent relationships, including those with group membership as a moderator, may lead to decision making that results in detrimental   individual,   group,   organizational,   and   societal consequences.

CONCLUSIONS

■ This chapter discussed the role of statistical inferences in testing moderating effects and framed statistical power within the larger issue of null hypothesis statistical significance testing. Then it provided a detailed discussion of a large number of factors that have a detrimental effect on the statistical power of MMR. These factors are related to variable distributions (i.e., predictor variable variance re-

duction, transformations of skewed criterion scores), operationalizations of criterion and predictor variables (i.e., measurement error,
scale coarseness, polichotomization of truly continuous variables),
sample size (i.e., total sample size, sample size across moderator-
based subgroups), and characteristics of the predictor variables (i.e.,
correlation between the predictor X and the moderator Z, correlation between the predictor X and the criterion Y). In one way or another, each of these factors has the potential to lower the sample-
based observed effect size as compared to the effect in the population. In turn, an underestimation of effect size causes power to drop.
Because a subset of these factors is present in most social science research situations, it is important to be aware of what they are and the
mechanisms through which they affect power. Moreover, the factors
that affect power downwardly also have interactive effects. This
means that it is sufficient to have an adverse value on just one or
two of the factors for the observed effect sizes to be substantially
lower than their population counterparts and for power to decrease
substantially below the suggested value of .80.

■ This chapter also showed that, unfortunately, most MMR users are
probably unaware of the dramatic impact that methodological and
statistical artifacts can have on power. A 30-year review of three major social science journals showed that the median effect size is only
.002. With such small observed effect sizes, it is no surprise that
moderators are hard to find (Chaplin, 1997); they were labeled "elusive" as early as in the 1970s (Zedeck, 1971). As long as such small
effects have a meaningful impact for science or practice within a
specific context, researchers should not rely on broad-based standards of what constitutes a critical effect size (e.g., Cohen, 1988).
Instead, researchers should conduct a power analysis and plan future research design based on smaller (and more realistic) targeted
effect sizes.

■ The small observed effect sizes, and consequently lower power, of
MMR has important implications for both theory and practice. Regarding theory, it is likely that low power has led to the incorrect
conclusion that there is no moderating effect in numerous research
domains. This implication can open whole new research avenues. It
may be useful to revisit published articles that reported no moderating effect in spite of well-grounded theoretical rationale and replicate the studies with MMR tests with greater power. We should not
be surprised if predicted moderator variables are now found, given
an appropriate power level. From a practical standpoint, low power
means that an intervention that does not have the same effect across

groups is nevertheless used uniformly. Such erroneous implementation of social science interventions (e.g., preemployment testing, clinical treatment, smoking cessation programs, and so forth) is likely to lead to uneven desired outcomes across groups (e.g., predicted job performance, mental health, smoking cessation success rate). Low power can lead to the unexpected result that a particular intervention did not work as well for all groups under consideration. Such a finding has implications in terms of fairness for the participants as well as the optimization of resources used in implementing the intervention.

■ But, in spite of all the bad news about power, there is hope that the power of MMR can reach adequate levels in many circumstances. Next, Chapter 6 describes strategies available to deal with each of the factors affecting power that were described in this chapter.

6

Light at the End of the Tunnel
How to Solve the Low-Power Problem

*Understand that most problems are a good sign. Problems
indicate that progress is being made, wheels are turning, you are
moving toward your goals. Beware when you have no problems.
Then you've really got a problem. . . . Problems are like
landmarks of progress.*
—SCOTT ALEXANDER

*Every problem has in it the seeds of its own solution. If you don't
have any problems, you don't get any seeds.*
—NORMAN VINCENT PEALE

Chapter 5 showed a fairly bleak picture of MMR. First, it showed
that there are numerous factors that lower the power of an MMR
analysis to such a level that it may be unlikely that a moderator will
be detected. Results of a literature review showed that observed effect sizes are quite smaller, even, than the traditional definition of
"small" effect (cf. Cohen, 1988), confirming the dramatic impact of
the factors that affect effect sizes and power as seen in published
research. In spite of this bad news, MMR users have some choices
in terms of how a study is designed and the data are collected such
that the adverse impact of the factors that lower power can be mini-
mized. This chapter provides a discussion of these remedies. The
discussion of the solutions is organized around each of the factors
described in Chapter 5 that have a detrimental effect on power.

Before these solutions are discussed, however, it is important to
make two caveats. First, as noted in Chapter 5, the factors that limit
or enhance power have interactive effects. Thus, although there is

no choice but to discuss each of them one by one, these factors should not be considered in isolation. Second, it may be the case that there are practical constraints that prevent a researcher from improving one of the factors (e.g., collecting further data and thereby increasing N), but there may be a possibility to improve another (e.g., decrease truncation on the predictor X). Such practical constraints lead to specific trade-offs in terms of power. This chapter does not address those trade-offs directly. However, Chapter 7 describes computer programs that allow MMR users to estimate power for a variety of possible research scenarios (e.g., increase of 10% in sample size and no improvement in reliability vs. increase of 5% in sample size and a .05 improvement in the reliability of the predictor X). Thus, the programs described in Chapter 7 allow MMR users to make informed decisions based on the cost and benefit of manipulating the factors under consideration.

HOW TO MINIMIZE THE IMPACT OF FACTORS AFFECTING THE POWER OF ALL INFERENTIAL TESTS

Sample Size

MMR users should strive to achieve a sample size that helps achieve a sufficient level of power. There are practical constraints in the data collection phase of any study that may preclude researchers from increasing the size of their samples. However, every effort must be made in the planning stage of the study, and before it is too late to collect further data, so that N is at an appropriate level.

Preset Type I Error Rate

Researchers usually fix the a priori probability of making a Type I error at a predefined value (e.g., .05). Because the power of MMR is inadequate in most research situations faced by social scientists, one way to enhance power is to increase the preset nominal Type I error rate above the traditional levels, for example to $\alpha = .10$ (Cascio & Zedeck, 1983). In other words, rather than fixing Type I error at a preset value and letting Type II error vary, this alternative strategy consists of determining the relative size of each. Although this is certainly a potentially good strategy to enhance power, MMR users may be reluctant to increase the preset Type I error rate for two reasons. First, increasing the preset Type I error rate may raise concerns among journal editors and reviewers. Second, any increase in the preset Type I error rate should be weighed against the potential disadvantage of increasing the possibility of finding a "false" moderator (i.e., committing a Type I error and incorrectly concluding that there is a moderator variable).

If one wishes to implement this strategy, one should make an informed decision about the specific trade-off between Type I and Type II errors rather than choosing an arbitrary value for alpha. Murphy and Myors (1998) suggested a useful way to weigh the pros and cons of increasing the Type I error rate for a specific research situation. The appropriate balance between Type I and Type II error rates can be achieved by using a preset Type I error rate that takes into account the "desired relative seriousness" (DRS) of making a Type I versus a Type II error. Because Type II error = 1 − power, this strategy is also useful for choosing an appropriate Type I error vis-à-vis statistical power.

Consider the situation where an MMR user is testing the hypothesis that the effectiveness of a training program for unemployed individuals varies by geography such that the training program is more effective in regions where the unemployment rate is higher than 6%. Assume that this researcher decides that the probability of making a Type II error or β (i.e., incorrectly concluding that unemployment rate in a region is not a moderator) should not be higher than .15. The researcher also decides that the seriousness of making a Type I error (i.e., incorrectly concluding that percentage of unemployment in a region is a moderator) is twice as serious as making a Type II error (i.e., DRS = 2). Assume the researcher makes the decision that DRS = 2 because a Type I error means that different versions of the training program would be needlessly developed for various regions, and this would represent a huge waste of resources. The desired preset Type I error can be computed as follows (Murphy & Myors, 1998):

$$\alpha_{desired} = \left[\frac{p(H_1)\beta}{1 - p(H_1)} \right] \left(\frac{1}{DRS} \right) \qquad (6.1)$$

where $p(H_1)$ is the estimated probability that the alternative hypothesis is true (i.e., there is a moderating effect), β is the Type II error rate, and DRS is the desired relative seriousness (i.e., a judgment of the seriousness of a Type I error vis-à-vis the seriousness of a Type II error).

For this example, assume that based on a strong theory-based rationale and previous experience with similar training programs, the MMR user estimates that $p(H_1) = .6$. Solving Equation 6.1 yields

$$\alpha_{desired} = \left[\frac{(.6)(.15)}{1 - .6} \right] \left(\frac{1}{2} \right) = .11$$

Thus, in this particular example, using a nominal Type I error rate of .11 would yield the desired level of balance between Type I and Type II statistical errors.

Moderating-Effect Size

MMR users typically do not have the ability to manipulate the size of the moderating effect in the population (i.e., β_3 or $\Delta\Psi^2$). After all, the purpose of the study is to learn whether the regression coefficient associated with the product term is different from zero and how big it is. However, using sound theory to make predictions—as opposed to going on "fishing expeditions" for moderating effects—is likely to affect the size of the moderating effect. As described in Chapter 1, there is no data analysis substitute for good thinking. An MMR analysis will have little power to detect an effect if the effect is too small in the population, or even nonexistent, because the substantive research question lacks sufficient conceptualization.

HOW TO MINIMIZE THE IMPACT OF FACTORS AFFECTING THE POWER OF MMR

Reduction of Variance in Predictor Variables

A difference in the variance of the predictor X between the population and the sample can be considered an instance of biased sampling. In the presence of truncation, which is one of several mechanisms leading to variance reduction, not all members in the population have the same probability of being selected in the sample. Therefore, the best way to avoid the truncation problem, which has a detrimental impact on power, is to draw random samples from the population.

In many situations encountered by social scientists, however, it is not possible to use randomly selected samples and avoid the truncation problem. For example, consider the situation where an MMR user wishes to test whether the relationship between preemployment test scores and job performance is similar for a sample of currently employed female and male workers. Most likely the goal of this study is to assess the fairness of the test in question (Cleary, 1968) so that it can be safely administered to job applicants in the future. However, the current employees were selected because their scores on the preemployment test were higher than a certain cutoff score. Therefore, the variance of scores for the preemployment test is smaller in the sample of employees as compared to the complete population of applicants. In this case, the power to detect a possible interaction effect between test scores and gender will be decreased by using the truncated sample of employees.

In addition to using randomly selected samples, what else can be done to address the variance reduction problem? McClelland and Judd (1993) showed that the use of the extreme-group design is another alternative. The extreme-group design involves sampling the highest and lowest scores of the distribution only and not sampling scores in the center of the distribution. The extreme-group design strategy also leads to a truncated distribution, but truncation takes place in the middle of the distribution. In contrast to the more typical truncation mechanism, the effect of implementing an extreme-group design is an increase in variance in the sample vis-à-vis the population and a subsequent increase in statistical power. That is, implementing an extreme-group design is an oversampling technique (Aiken & West, 1993) that results in sample scores suffering from variance enhancement (Hunter & Schmidt, 1990) as compared to the population scores. Although the extreme-group design provides the advantage of greater statistical power, it is at the expense of artificially increasing the sample variance (Cortina & DeShon, 1998). Range enhancement does not produce a bias in MMR's statistical significance test. However, it does increase the estimated effect size and biases the comparison of relative strength of the moderating effect vis-à-vis the other predictors in the MMR equation (Bobko & Russell, 1994; Cortina & DeShon, 1998). In short, the use of an extreme-group design increases sample variance and, consequently, increases statistical power. However, results regarding the size of the moderating effect should be interpreted with caution because they are based on a (range-enhanced) biased sample rather than on a random sample (Aguinis, 1995).

Transformations of Skewed Criterion Scores

As noted in Chapter 5, transformations of a positively skewed criterion score distribution such as a logarithmic function yield smaller moderating effect estimates as compared to implementing a simple percentile bootstrap procedure and then performing a traditional OLS regression analysis (Russell & Dean, 2000). Bootstrapping consists of simulating the process of randomly selecting samples from a population in order to construct a sampling distribution (Diaconis & Efron, 1983; Efron, 1979). This computer-intensive method consists of copying scores on Y, X, and Z from one sample of size N thousands of times and storing the resulting samples in separate files. Statistics then can be computed from each sample and a sampling distribution of the statistic can be created.

Each newly created sample m of size N includes various combinations of scores from the original sample (e.g., one individual's scores Y_1, X_1, and Z_1 may be represented 10 times in m_1 and 20 times in m_2). The percentile bootstrap method used by Russell and Dean (2000; see Efron & Tibshirani, 1993, Chapter 13) involves the following steps:

1. Create $m = 1,000$ bootstrap samples including scores for Y, X, and Z. There are several software packages that allow for bootstrap estimates (e.g., SYSTAT v. 10.2, *http://www.systat.com/products/Systat/*).
2. Follow all steps described in Chapter 2 regarding how to conduct an MMR analysis with each of the samples. This can be a very time-consuming task. Therefore, it would be useful to use a program that reads the data included in each of the samples, computes the product term, and computes statistics including ΔR^2 (Russell & Dean, 2000, will provide a copy of their program upon request).
3. Rank order the ΔR^2s obtained from each of the 1,000 samples.
4. The 25th and 975th values for ΔR^2 are used as the upper and lower limits of the 95% confidence interval.

In sum, the advice regarding how to deal with this problem is simple: When faced with a positively skewed criterion, do not log-transform your data; instead, implement a bootstrap procedure on the skewed (untransformed) data.

Measurement Error

There is no shortcut to dealing with the measurement error problem: MMR users must be aware of the detrimental impact of measurement error in X on power and must strive to develop and use reliable measures (Cascio & Aguinis, 2003; Evans, 1991a). As noted in Chapter 5, there is a compounded impact of the measurement error in the predictor X and the moderator Z on the measurement error of the $X \cdot Z$ product term (e.g., Busemeyer & Jones, 1983). Thus, even a .05 improvement in the reliability of one of the predictors can have a noticeable effect on the reliability of the product term.

Scale Coarseness

The discrepancy in the number of anchors in the criterion variable and the $X \cdot Z$ product term reduces power, particularly when the criterion scale does not contain at least as many response options as exist in the theoretical response domain (Russell & Bobko, 1992). A way to minimize this potential problem is to use a continuous or nearly continuous criterion scale. Reassuringly, increasing the number of response categories does not have an attenuating effect on reliability (Cicchetti, Showalter, & Tyrer 1985). For example, reliability results reported by Russell and Bobko (1992) indicated no difference in random measurement error between a traditional Likert-type and a continuous scale format.

A way to implement the suggested strategy is to record responses to paper-and-pencil Likert-type questions on a graphic line segment and then measure the distance between a respondent's mark and the end of the line segment manually (e.g., Arnold, 1981; Russell & Bobko, 1992). However, this procedure is not necessarily a good practical suggestion because it is time-consuming and prone to errors. Fortunately, a computer program that overcomes these shortcomings is available (Aguinis, Bommer, & Pierce, 1996). The Computer Administered Questionnaire (CAQ) program was written in Turbo C and administers questionnaires on IBM and IBM-compatible personal computers by prompting respondents to indicate their answer by clicking on a graphic line segment displayed on the screen. The CAQ program includes easily modifiable anchors for the graphic line segment (e.g., agree–disagree, satisfied–dissatisfied), no limit to the number of questions that can be included in the questionnaire, and responses that are stored directly into an ASCII file, thereby eliminating the need to perform any data entry work before conducting the subsequent MMR analysis. Responses provided on the line graph segment may range from the utmost left extreme of the scale (i.e., 000000, indicating 0% of the line) to the utmost right extreme (i.e., 100000, indicating 100.000% of the line), in increments of approximately 000160 (i.e., .16%) on EGA/VGA monitors, and 000318 (i.e., .32%) on CGA monitors.

To use the CAQ program, the researcher first needs to create an ASCII file that includes the instructions, questions, debriefing statement, and other statements to be displayed on the screen. Respondents then provide their answers to each of the questions by using the mouse and clicking on any point on the continuum or, for user-defined questions about demographic and other background information, by using numbers on the keyboard. Respondents advance through the questions sequentially by clicking on a NEXT icon displayed at the bottom of each screen.

The first question in the program, which is a default that can be altered by the researcher, inquires about the last four digits of the respondent's Social Security number as a method for identifying unique cases during subsequent data analyses. Then a practice screen allows respondents to become comfortable using the mouse to provide their answers on the line displayed on the computer screen. This screen is included in Figure 6.1.

Finally, an additional feature of CAQ is that participants' responses are appended to the existing output file. Thus, the same floppy diskette can be used for as many respondents as the available disk space permits. This feature can be especially helpful if a researcher is implementing a snowballing sampling methodology whereby an individual can give the floppy diskette to another one once he or she has participated in the study. The program is available at *www.cudenver.edu/~haguinis/mmr*.

Polichotomization of Truly Continuous Variables

The problem of dichotomizing or polichotomizing truly continuous variables occurs when researchers implement procedures such as a "median split" on truly continuous variables (e.g., age is artificially dichotomized into Group 1 = under 40 and Group 2 = 40 and over). Traditionally, polichotomization has been used so that researchers would be able to conduct an ANOVA using the artificially created groups. But with the advent of multiple regression, researchers should avoid dichotomizing or polichotomizing truly continuous variables (Stone, 1988). This procedure has the effect of reducing the variance of the variables involved, leading in many cases to a subsequent decrease in power.

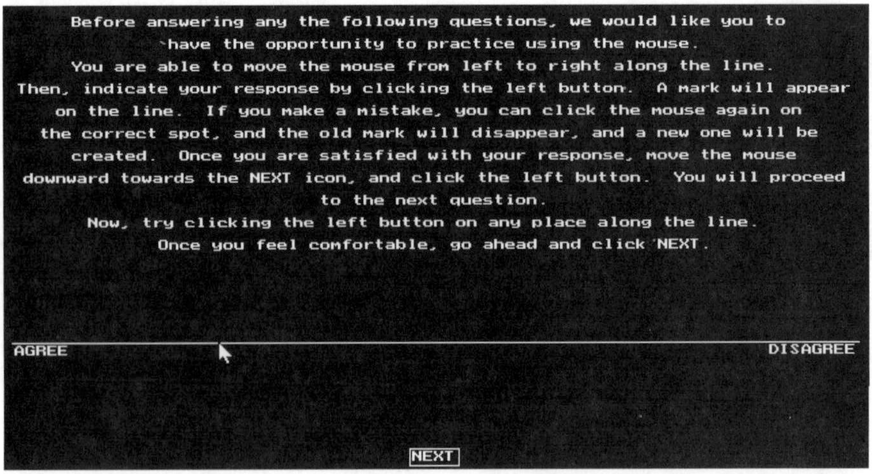

FIGURE 6.1. Practice screen from the CAQ program (Aguinis, Bommer, & Pierce, 1996).

Total Sample Size

This issue was discussed earlier in the section addressing factors that affect the power of all inferential tests.

Sample Size across Moderator-Based Subgroups

Because unequal sample sizes across moderator-based subgroups reduce power, equalizing the numbers by oversampling from the smaller group is a strategy that will increase power. This strategy is similar to the recommendation regarding the use of an extreme-group design to deal with truncation problems, and the same caveats apply. For example, equalizing the number of African Americans and Whites in one's sample, given that the sizes of these groups in the population are not equal, increases the power of the MMR test at the expense of a nonrepresentative sample.

One can think of sample sizes across the moderator-based subgroups as being closely related conceptually to the variance of a continuous variable. However, sample size across the moderator-based subgroups is not the same as the variance of the categorical moderator Z. For example, the variance of a binary moderator Z is (Aguinis, Boik, & Pierce, 2001) as follows:

$$s_z^2 = \frac{\Sigma(Z_i - \bar{Z})^2}{N-1} = \frac{Np(1-p)}{N-1} \tag{6.2}$$

where $N = n_1$ (i.e., sample size in Subgroup 1) + n_2 (i.e., sample size in Subgroup 2), and $p = n_1/N$. Note that an increase in sample size can result in a decrease or an increase in the variance of Z. For example, if $n_1 = 10$ and $n_2 = 40$, then the sample variance is $8/49 = .1636$. If n_1 is increased to 40, then the sample variance is $20/79 = .253$ (i.e., an increase). If n_1 remains at 10 and n_2 is increased to 50, then the sample variance is $25/177 = .1412$ (i.e., a decrease) (Aguinis, Boik, & Pierce, 2001).

Practical considerations must also be taken into account in sampling from the various groups. Assume that we are interested in a binary moderator variable and that the cost for sampling from Group 1 is C_1 whereas the cost for sampling from Group 2 is C_2. Further, assume that the total amount of money available for this research project is C so that $(n_1)(C_1) = (n_2)(C_2)$. To maximize power it is optimal to choose

$$\frac{n_1}{n_2} = \sqrt{\frac{C_1}{C_2}}$$

so when $C_1 = C_2$, then $n_1 = n_2$.

In short, we can address the problem of unequal sample sizes by oversampling from the smaller group and equalizing the groups in the sample. However, this solution results in nonrepresentative samples; although the statistical significance of MMR's F test is not biased, the estimate of the size of the moderating effect is.

Correlation between the Predictor X and the Moderator Z

As noted in Chapter 5, some have argued that the presence of multicollinearity (i.e., correlation between the predictor X and the moderator Z) leads to an ill-conditioned solution in which the regression coefficients are unstable, error terms are increased, and the power of MMR may be consequently decreased. However, the concerns about the detrimental effects of multicollinearity on MMR power have proven to be unfounded (Cronbach, 1987).

Correlation between the Predictor X and the Criterion Y

To be able to detect the moderating effect of Z it is generally helpful to have a strong X–Y relationship in the first place (Rogers, 2002). This situation applies particularly to ordinal effects (i.e., when the lines regressing Y and X for the various moderator-based subgroups do not cross within the possible range of the values of X). For example, Evans (1985) found that when the product term explained 1% of variance in Y over and above the first-order effects (i.e., $\Delta R^2 = .01$), this effect was statistically significant if the R^2 for the first-order effects was approximately .80. However, the same $\Delta R^2 = .01$ was not statistically significant when the R^2 for the first-order effects was approximately .30.

Once again, the role of theory is critical in identifying a predictor variable X that is indeed related to the criterion Y. This point is not related to measurement or data analysis, and the "garbage in, garbage out" motto applies: An ill-conceived hypothesis that is not well grounded in theory or past research is possibly the worst enemy of statistical power. There are no measurement and statistical patch-ups for an ill-conceived study that does not ask the right theory-related questions.

CONCLUSIONS

■ Although there are numerous factors that affect the power of MMR adversely, MMR users can make design and measurement choices that can enhance statistical power. This chapter described 16 strategies available to address each of the methodological and statistical factors discussed in Chapter 5.

■ A summary of these recommendations is presented in Table 6.1. Not all strategies are available in every research situation. However, implementing as many of the available strategies as possible is an important consideration that should be taken seriously by all MMR users who wish to minimize the impact of the factors that are likely to lead to the incorrect conclusion that there is no moderating effect. Furthermore, researchers should not wait until the data have been collected to think about statistical power. On the contrary, the implementation of strategies that enhance power should begin at the research planning and design stages of conducting any study. Next, Chapter 7 describes computerized tools that allow for the computation of statistical power given specific study characteristics (e.g., sample size, measurement error, and so forth).

TABLE 6.1. Summary of Recommended Strategies to Minimize the Adverse Effects of Factors Affecting Power as Described in Chapter 5

Factor affecting power	Strategy to increase power
Total sample size	• Plan research design so that sample size is sufficiently large to detect the expected effect size. • Compute power under various sample-size scenarios using programs described in Chapter 7 so that sample size is not unnecessarily large, thereby causing an unnecessary expense in terms of time and money.
Preset Type I error	• Do not feel obligated to use the conventional .05 level. • Use a preset Type I error based on the judgment of the seriousness of a Type I error vis-à-vis the seriousness of a Type II error.
Moderating effect size	• Use sound theory to make predictions about moderating effects as opposed to going on "fishing expeditions." • Make sure the substantive research question has a sufficient level of conceptualization; an MMR analysis will have little power to detect an effect if the effect is too small or even nonexistent.
Predictor variable variance reduction	• Draw random samples from the population. • Use an extreme-group design (but remember that this is done at the expense of increasing the sample variance).
Transformations of skewed criterion scores	• Do not log-transform skewed criterion scores. • Implement a bootstrap procedure using the original (untransformed) skewed data.
Measurement error	• Develop and use reliable measures.
Scale coarseness	• Use a continuous criterion scale; this can be done by recording responses on a graphic line segment and then measuring them manually, or by using CAQ (available at *www.cudenver.edu/~haguinis/mmr*) or other programs that prompt respondents to indicate their answer by clicking on a graphic line segment displayed on the screen.
Polichotomization of truly continuous variables	• Do not dichotomize or polichotomize truly continuous variables.
Sample size across moderator-based subgroups	• Equalize the sample sizes across subgroups by oversampling from the smaller groups (but remember that this is done at the expense of a resulting nonrepresentative sample).
Correlation between the predictor X and the moderator Z	• The concern that predictor intercorrelation has a detrimental impact on the statistical power of MMR is unfounded.
Correlation between the predictor X and the criterion Y	• Use sound theory to identify a predictor that is strongly related to the criterion because the correlation between the predictor X and the criterion Y is positively related to statistical power.

7

Computing Statistical Power

I think knowing what you cannot do is more important than knowing what you can.
— LUCILLE BALL

It is not good to know more unless we do more with what we already know.
— R. K. BERGETHON

Chapter 5 described the adverse effects of several factors on the statistical power of MMR, and Chapter 6 discussed suggested strategies to address each of the factors that lower power. This chapter discusses reasons why it is useful to compute statistical power, then it includes a description and illustrations of two empirically based and one theory-based program(s) that allow for the computation of the power of an MMR test. Finally, the chapter includes a discussion of the relative impact of the various factors that affect power.

USEFULNESS OF COMPUTING STATISTICAL POWER

As a reminder, statistical power is the probability of detecting a moderating effect in the sample when the moderating effect exists in the population. There are many reasons why it is useful to compute power. First, given the numerous methodological and statistical factors that have an adverse impact on power, it is particularly important that MMR users compute power *before* testing their moderating-effect hypotheses. If a researcher knows in advance that the MMR test has sufficient power, then null results regarding the moderating effect can be interpreted with confidence. On the other hand, a low-power MMR test resulting in a null finding is inconclusive.

For example, assume a power analysis results in a value of .90 (i.e., there is a 90% chance that the population moderating effect will be detected in the sample). A researcher can trust the null finding regarding a moderating effect because, given a .90 level of power, it is in fact very likely that there is no moderating effect. Alternatively, assume a power analysis results in a value of .20 and the MMR analysis also indicates no moderating effect. How do we know that this is not the result of a Type II error? In other words, how do we know that in fact the moderating effect does exist in the population, but our MMR test did not detect it? Given low power, it is not known whether the null finding is the result of a Type II error. Of course, no researcher wishes to spend months, or years, planning and conducting a study and then reach the conclusion that he or she is unable to test the hypothesis about the moderating effect with any degree of confidence and, therefore, results are inconclusive. In short, the first reason why the computation of power is useful is that it allows researchers to anticipate what the resulting power will be, given a specific research situation, in which sample size, possible predictor variable variance reduction, and all the other factors that have the potential to affect power adversely may be present. If results of this power analysis show an insufficient level of power, then the study should not be conducted because null results are likely to be inconclusive.

The second reason why the computation of power is useful is that if the anticipated power value is low, then a researcher can conduct a cost–benefit analysis of the various alternatives he or she has available to enhance power. For example, one can compute the power resulting from increasing sample size by 20% as compared to increasing the reliability of the predictor scores by increasing the instrument's length by 30%. Given the cost associated with an increase in sample size vis-à-vis the improvement in the measurement instrument's reliability, which of these strategies would be more cost-effective in terms of improving power? In short, being able to compute the anticipated power for several alternative design and measurement scenarios allows MMR users to make better use of their resources in attempting to maximize power.

The third reason why computing power is useful is that researchers are able to revisit past null findings regarding moderating effects. That is, the tools that are described in this chapter will allow researchers to go back to published studies showing the "absence" of a moderating effect and check whether the study was conducted at an insufficient power level. If this is the case, it would be useful to replicate such studies with a sufficient power level to conduct a more accurate test of the moderating-effect hypothesis. This type of retrospective power analysis should not be confused with a post hoc power analysis that uses the observed effect size as the target effect size. A post hoc power analysis pro-

ongoing

An effect size describes how large the relationship is between two variables vides no additional information beyond that contained in the confidence interval (CI) around the parameter estimate of interest. Thus, the target effect size used in a retrospective power analysis should not be based on the observed effect size but on a specific "critical" effect size. The magnitude of this "critical" effect size depends on the research area, or even more particularly on the specific content and research methods being employed (Cohen, 1977) as well as the nature of the outcome studied (Eden, 2002).

The fourth reason why the computation of power is useful is that many researchers may find that their hypotheses about moderating effects have not been supported and may wonder if this is a result of inadequate statistical power. Although it is always advisable to check statistical power *before* a study is conducted, it may also be useful to check statistical power after a study has been conducted, particularly when results show a null finding regarding a moderating effect. In these situations of null findings and low power, MMR users are presented with two possible conclusions:

- *Conclusion 1*: The population moderating effect equals zero (i.e., there is no moderating effect), or
- *Conclusion 2*: The population moderating effect is not zero but it has not been detected due to low statistical power.

If statistical power is inadequate, then Conclusion 1 must be taken with great caution: We just do not know if this is correct. There may or may not be a moderating effect in the population. However, it is better to know that the null finding may be due to inadequate statistical power than to make the incorrect conclusion, and put it in print, that the hypothesis about the presence of a moderating effect is incorrect. In short, it is better to acknowledge a study's weakness and suggest the need for future replication than to let oneself, and others, conclude perhaps incorrectly that there is no moderating effect. Finally, as noted earlier, this type of retrospective power analysis does not use the observed effect size as the target effect size. Instead, the target effect size is a critical effect size that is determined based on the specific research context and outcomes.

The fifth reason why the computation of power is useful is that funding agencies do not want to support research that will lead to inconclusive results. Of course, this is precisely what a low-power study does. Thus, many funding agencies demand that MMR users report a power analysis before a research program is given monies. For example, The Robert Wood Johnson Foundation's Web page includes a number of suggestions for strengthening grant proposals (*http://www.saprp.org/*).

One of the recommendations is that researchers include results of a power analysis. In short, computing power is often necessary in submitting proposals for research funding.

The sixth reason why computing power is useful is that journal editors and reviewers are becoming more aware of the weaknesses of null hypothesis significance testing and the need to address statistical power issues (Task Force on Statistical Inference, 2000). Thus, many journal editors and reviewers request results of a power analysis, particularly when results indicate that the null hypothesis should not be rejected.

Next is a description of three programs that allow for the computation of the power of MMR. There are two types of programs: empirically based and theory-based. The empirically based programs were designed using results of Monte Carlo simulations. The theory-based program was designed based on analytic work. Each of these programs, which are available at *www.cudenver.edu/~haguinis/mmr*, is described next.

EMPIRICALLY BASED PROGRAMS

Program POWER

The program POWER allows for the computation of power for situations involving a binary moderator variable (Aguinis, Pierce, & Stone-Romero, 1994). This was the first program to compute the power of MMR whose description was published in a refereed journal.

Specifications

POWER was originally written in QuickBASIC (release 4.5) (Aguinis et al., 1994). However, the program has been translated into Java. Java is an increasingly popular programming language because of its flexibility for World Wide Web applications. Thus, in contrast to the original version described by Aguinis et al. (1994), POWER can now be executed remotely on the Web by using any Web browser (e.g., Internet Explorer, Netscape Navigator). Thus, POWER can be executed regardless of operating system platform (e.g., Windows 98/NT/2000/XP, Macintosh, OS2).

POWER is based on empirical results obtained by Stone-Romero et al. (1994), who conducted a simulation to investigate the effects on power of total sample size, sample size across moderator-based subgroups, and the difference between the X–Y correlation across the two categories of the moderator Z. After their simulation was completed, Stone-Romero et al. (1994) conducted a multiple regression analysis in which the criterion was the empirically obtained statistical power values (i.e., proportion of times that the null hypothesis regarding the existence of a moderator variable was correctly rejected), and the predictors were the variables representing the first-order and interactive effects of the manipulated variables (i.e., differences in subgroup sample sizes, total sample size, and difference between X–Y correlation coefficients). Note that in Stone-Romero et al.'s simulation, the magnitude of the moderating effect in the population was operationally defined as the absolute difference between the X–Y correlations across the two moderator-variable-based subgroups. This operationalization for moderating effect size was possible because the predictor X and the criterion Y were generated as normal deviates (i.e., mean of 0 and SD of 1). In other words, because $b = r\left(s_Y / s_X\right)$ and in the simulation $s_Y = s_X = 1$, differences between correlation coefficients are equivalent to differences between regression coefficients. POWER uses the prediction equation from the Stone-Romero et al. (1994) regression equation.

regr-coefficient   *correlation coef.*

Example

To run the program, simply visit *www.cudenver.edu/~haguinis/mmr* and click on the "Run POWER" icon. The program POWER will be executed and the Java Applet window reproduced in Figure 7.1 will appear on your screen.

Assume you are at the planning stages of a study investigating a hypothesized effect of a binary moderator and wish to compute power. Based on previous research and theory considerations, you believe the correlation between X and Y is .20 for Group 1 and the correlation between X and Y is .30 for Group 2. Moreover, you feel that based on the resources available for this study, you will be able to collect data from 50 participants in Group 1 and from 100 participants in Group 2.

The program POWER prompts users interactively to input the sample size in each of the groups and the sample-based correlation coefficient between X and Y for each of the groups. After the user inputs these values, POWER displays the estimated power to detect a moderating effect for the specified conditions. Figure 7.2 reproduces the output screen. In this case, the power of the MMR test is .104. That is, there is only a 10% chance that given these conditions the researcher would be able to detect an existing moderating effect in this particular illustrative study. Obviously, this is a very low value given that the power ranges from 0 to 1.00 (Casella & Berger, 2002) and its recommended minimum value is .80 (Cohen, 1988).

Limitations

The program POWER is useful for estimating the power to detect binary moderating effect, but it has the following limitations:

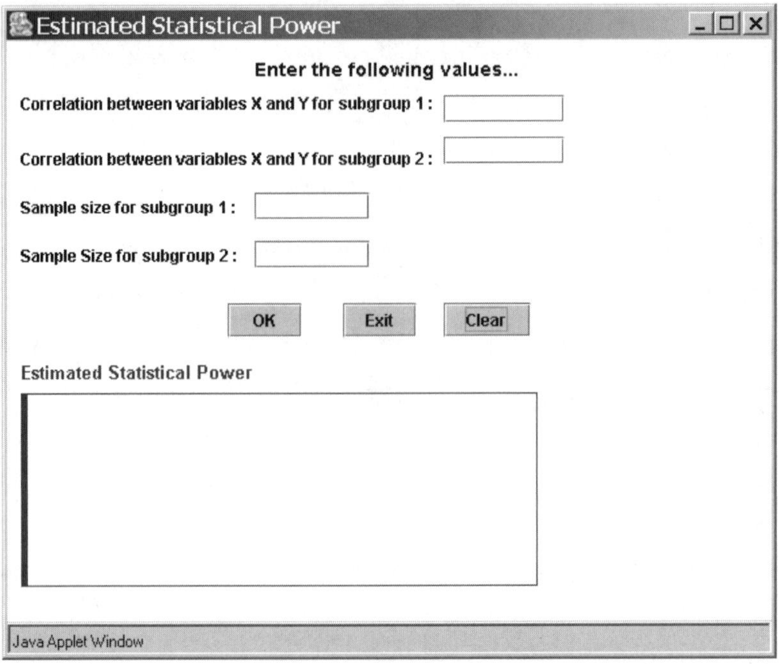

FIGURE 7.1. Java Applet window for the program POWER.

- It allows for the computation of power for tests involving binary moderators only.
- It does not take into account the effect of several factors known to affect power (e.g., predictor variable variance reduction, measurement error).
- Its accuracy may be limited to the value ranges included in the Monte Carlo studies upon which it is based (i.e., Stone-Romero et al., 1994). For example, Stone-Romero et al. used values for total sample size of 60 and 300. Thus, the program POWER might not provide accurate power estimates for situations having sample size values far from 60 or 300.
- It assumes that the ratio s_Y/s_X is constant across groups and, therefore, it assumes that the difference between correlation coefficients across groups is equivalent to the difference between regression coefficients across groups.
- It does not include the interactive effects of various methodological and statistical artifacts on power.
- It assumes the relationship between sample size, unequal sample

size across moderator-based subgroups, and effect size with statistical power is linear.

In short, POWER is easy to use, it is fast, and it does not require large computer resources, but it also has some limitations. Overall, however, it provides a good "quick and dirty" power value for tests involving binary moderator variables.

Program MMRPWR

The program MMRPWR also allows for the computation of power for tests including binary moderators only, but it includes several improvements over the program POWER (Aguinis & Pierce, 1998b).

Specifications

MMRPWR was originally written in QuickBASIC (release 4.5) (Aguinis & Pierce, 1998b). However, as in the case of the program POWER,

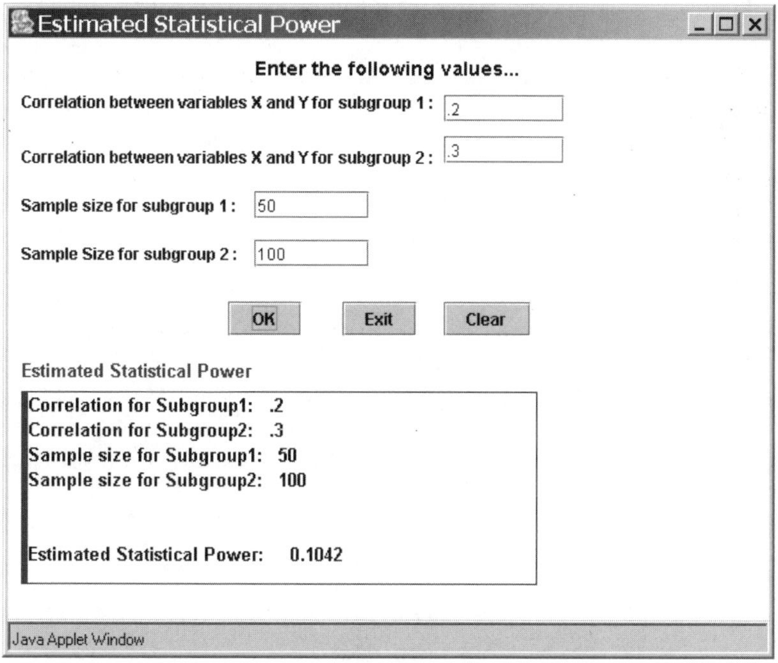

Estimated Statistical Power

Enter the following values...

Correlation between variables X and Y for subgroup 1 : `.2`

Correlation between variables X and Y for subgroup 2 : `.3`

Sample size for subgroup 1 : `50`

Sample Size for subgroup 2 : `100`

[OK]   [Exit]   [Clear]

Estimated Statistical Power

Correlation for Subgroup1: .2
Correlation for Subgroup2: .3
Sample size for Subgroup1: 50
Sample size for Subgroup2: 100

Estimated Statistical Power: 0.1042

Java Applet Window

FIGURE 7.2. Illustrative output screen for the program POWER.

MMRPWR has been translated into Java. Thus, MMRPWR can now be executed remotely on the Web by using any Internet browser and any operating system platform.

MMRPWR is based on empirical results obtained by Aguinis and Stone-Romero (1997), who conducted a more extended simulation than Stone-Romero et al. (1994) did. The program MMRPWR uses the prediction equation that includes empirically obtained power value as the criterion and the manipulated methodological and statistical artifacts as predictors. Specifically, these artifacts are the following:

- Discrepancy between sample and population X variance (i.e., X standard deviation in sample/X standard deviation in population).
- Total sample size (N).
- Proportion of cases in moderator-based subgroup 1 (i.e., $p_1 = n_1/N$).
- Fisher's z transformation of the sample-based correlation between the predictor X and the moderator Z.
- Absolute value of the difference between the Fisher's z transformations of the sample-based correlations between X and Y for Subgroup 1 and Subgroup 2.

Similar to Stone-Romero et al. (1994), Aguinis and Stone-Romero (1997) operationalized the moderating effect size in the population using X–Y correlations and $s_Y = s_X = 1$ across the two moderator variable-based subgroups so that differences between correlation coefficients are equivalent to differences between regression coefficients.

The prediction equation used by the program MMRPWR utilizes the proportion of scores in the smallest group relative to total sample size as opposed to the actual n_1 and n_2 values used as input. Then, after the power value is computed, the program implements an inverse of the arcs in square root transformation to the obtained power estimate. This transformation is needed because it is the inverse of the transformation used by Aguinis and Stone-Romero (1997) to linearize the power function (Winer, Brown, & Michels, 1991, p. 356, case ii). Thus, MMRPWR's output produces the estimated power expressed in the typical proportion metric. Finally, MMRPWR also computes a 95% CI around the estimated power value.

Example

To run the program, simply visit *www.cudenver.edu/~haguinis/mmr* and click on the "Run MMRPWR" icon. The program MMRPOWER will be executed and the Java Applet window reproduced in Figure 7.3 will appear on your screen.

We can compute power for the same scenario that we used above for the program POWER. That is, assume $r_{XY(1)} = .20$, $n_1 = 50$, $r_{XY(2)} = .30$, and $n_2 = 100$. Recall that the group with the smallest n should be entered as Subgroup 1. In addition, the program MMRPWR requests input regarding the X–Z correlation (i.e., r_{XZ}), and the standard deviation for X in both the sample and the population (note that SD_X is assumed to be constant across groups). For this illustration, assume a medium-size correlation between X and Z (i.e., $r_{XZ} = .30$), and no difference between the sample and the population X standard deviations (i.e., $SD_X = \sigma_X = 1$).

Not surprisingly, given that we used similar input data, Figure 7.4 shows that the resulting power value (i.e., .1039) is virtually identical to the one obtained using the program POWER (i.e., .1042). MMRPWR also provides a 95% CI around the power estimate. Figure 7.4 shows that the interval goes from .021 to .239.

Now, let us compute power using MMRPWR for the same scenario, but creating a difference between the sample and population X variance such that $SD_X = 1$ and $\sigma_X = 1.5$. Figure 7.5 shows the output screen for this new scenario. Although power was already low for the situation involving no difference between the population and sample variances for X, introducing a variance difference lowers it further to .086.

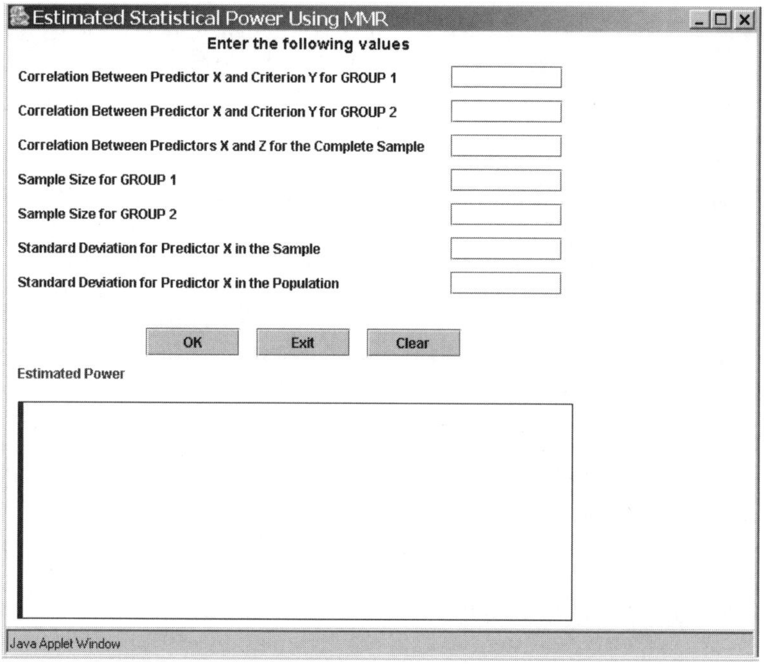

FIGURE 7.3. Java Applet window for the program MMRPWR.

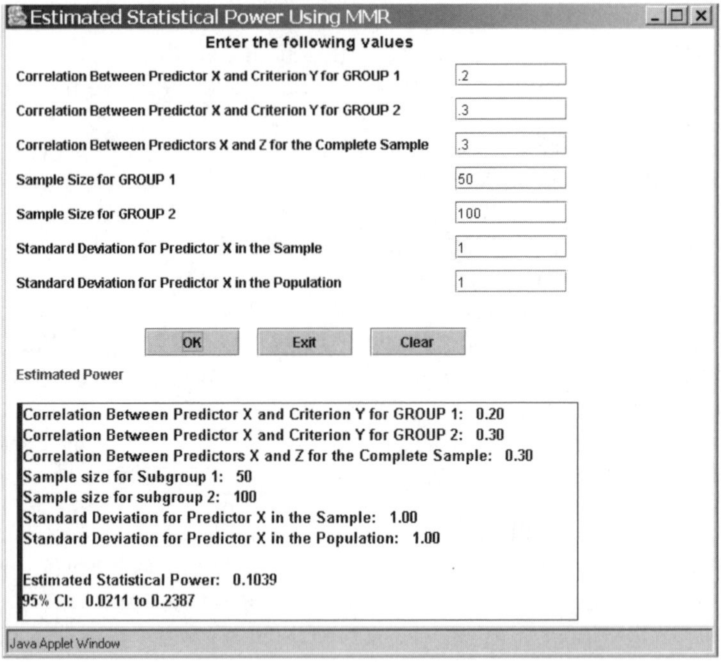

FIGURE 7.4. Illustrative output screen for the program MMRPWR assuming no difference between sample and population X scores variance.

Limitations

The program MMRPWR overcomes several of the limitations of the program POWER. Specifically, it does not assume a linear relationship between power and the various factors that affect power; it considers the impact of $SD_X < \sigma_X$, provides a confidence interval around the power point estimate, and considers the interactive effects on power of several methodological and statistical artifacts. However, this program does have the following limitations:

- Power is computed based on a preset Type I error (α) of .05 only.
- It may be difficult to obtain a good estimate for σ_X; also, restriction on X is assumed to take on only the simplest form of truncation.
- It does not include additional factors known to affect the power of MMR (e.g., measurement error).
- It allows for the computation of power for tests involving binary moderators only.
- Its accuracy may be limited to the value ranges included in the Monte Carlo studies upon which it is based (i.e., Aguinis &

Stone-Romero, 1997); for example, Aguinis and Stone-Romero (1997) did not include conditions for which the X–Y relationship was negative for one of the moderator-based subgroups.

- It assumes that the ratio s_Y/s_X is constant across groups, and, therefore, it assumes that the difference between correlation coefficients across groups is equivalent to the difference between regression coefficients across groups.

In short, MMRPWR overcame several limitations of the program POWER. It allows for additional information to be used in computing power and, therefore, it can be used in a more diverse set of situations.

THEORY-BASED PROGRAM

Program MMRPOWER

In contrast to POWER and MMRPWR, the program MMRPOWER is not based on an empirically derived regression equation linking meth-

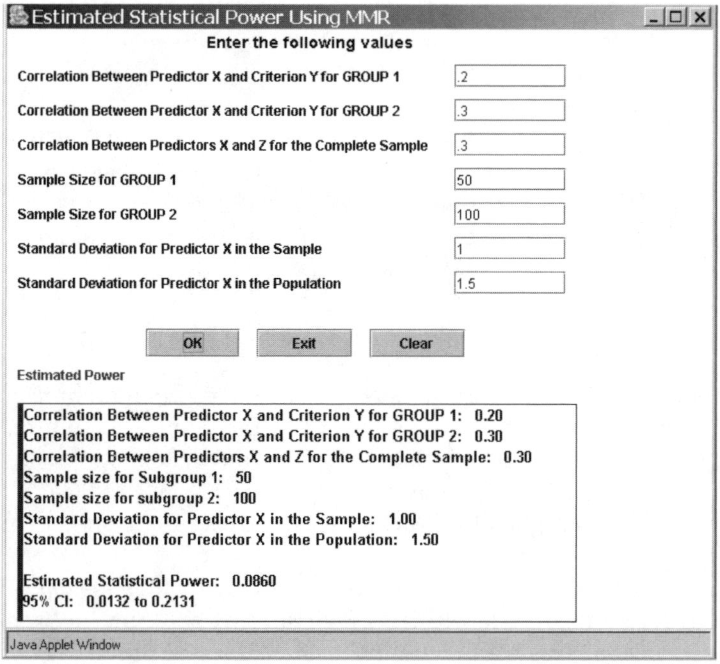

FIGURE 7.5. Illustrative output screen for the program MMRPWR assuming a difference between sample and population X scores variance.

odological and statistical artifacts with power. Instead, it is based on a theory-derived power algorithm developed by Aguinis, Boik, and Pierce (2001). This theory-based solution for approximating power is included in Appendix E. Although the presentation in Appendix E is fairly technical, the theorem shows that the power of the MMR F test depends on the following 10 quantities:

1. Preset nominal Type I error rate.
2. Number of moderator-based subpopulations.
3. Sample sizes across moderator-based subgroups.
4. Difference in slopes of Y on X across moderator-based subpopulations.
5. Reliabilities for Y in each of the moderator-based subpopulations.
6. Reliabilities for X in each of the moderator-based subpopulations.
7. Correlation coefficients between X and Y in each of the moderator-based subpopulations.
8. Residual variance of Y in each of the moderator-based subpopulations.
9. Variance of X in each of the moderator-based subpopulations.
10. Ratio of expected sample variance of X divided by population variance of X.

Note that some of the preceding 10 factors are not independent (Aguinis, Boik, & Pierce, 2001). For example, slopes change as a function of reliabilities, correlations, residual variances of Y, and variances of X. Nevertheless, the theory-based algorithm synthesizes research reviewed in Chapter 5 regarding the variables that affect the power of MMR.

Specifications

The program MMRPOWER performs all necessary computations to obtain a power approximation based on the theory-based solution. The program was written in FORTRAN 77 and is executable by using any Internet browser. Thus, similar to the updated versions of POWER and MMRPWR, the program MMRPOWER can be executed remotely on the Web.

To run the program, simply visit *www.cudenver.edu/~haguinis/mmr* and click on the "Run MMRPOWER" icon. The first screen of the program, shown in Figure 7.6, prompts for the following input: (1) number of moderator-based subgroups (the maximum number handled by the program is 20); (2) preset Type I error rate; (3) desired test (i.e., overall test of equality of slopes across moderator-based subgroups or tests of specific contrasts of slopes); and (4) sampling restrictions (i.e., none, sampling from truncated X normal distributions, or sampling from non-normal X distributions). The program also prompts the user to provide information regarding whether the input format includes correlations or slopes and whether they were based on true (i.e., population and error-free) or observable (i.e., sample-based and affected by measurement error) scores.

Once the information has been input on the first screen, a second screen appears prompting users to input the necessary information to compute power. In addition to sample size and reliabilities for each of the moderator-based subgroups, the necessary input varies depending on the choices made on the first screen. That is, one needs to input correlations or slopes based on true or observable scores for each moderator-based subgroup. Typically, one uses sample-based (i.e., fallible observed) scores as opposed to error-free true scores because it is expected that the data to be collected will not be error-free. In addition, users are prompted for (1) truncation proportion for X (i.e., proportion of scores that cannot be included in the sample) if truncated normal distributions were noted on the

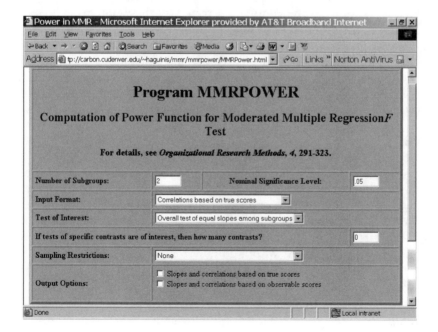

FIGURE 7.6. Initial screen for the program MMRPOWER.

first screen, (2) no information if no sampling restrictions were noted on the first screen, and (3) variance multiplying factors (i.e., expected sample variance/population variance) if non-normal X distributions were noted on the first screen. Finally, if tests of specific contrasts were requested on the first screen, the second screen prompts the user for contrast coefficients (Chapter 8 discusses contrast coefficients for situations involving more than two moderator-based subgroups).

Example

The opening screen is included in Figure 7.6. Let us use the same illustrative data we have used for the previous two programs. Thus, the situation includes two subgroups and a preset Type I error rate of .05. Also, the input choice is correlations based on observed scores (of course, we could also input slopes based on observed scores). Because there are two groups only there is no need to select specific contrasts, so we can select overall test of equality of slopes across subgroups. As for the case of the program MMRPWR as well, we can select a sampling restriction such that the X scores are normally distributed and are truncated. Finally, for the output options, let us choose slopes and correlations based on observable scores because we are using observable scores as input. Figure 7.7 includes the computer screen with these options.

Next, the program MMRPOWER shows a screen in which we need to input the data to compute power. To be consistent, let us use the same data we used before for sample size and correlations in each of the two moderator-based subgroups. In addition, assume the value of 1 for the X and Y standard deviations for both subgroups and a similar truncation proportion of .40 (i.e., 40% of the population scores are not included in the sample) across subgroups. Finally, assume that reliability values for X and Y are what is typically considered adequate (i.e., .80) for both subgroups. Figure 7.8 shows the resulting input screen.

Figure 7.9 shows the output screen for the values we entered on the two input screens. The program MMRPWR described in the previous section assumes that variables are measured with perfect reliability. Thus, not surprisingly, the resulting power value obtained with reliabilities of .80 is lowered to .069. The output screen also includes information regarding the within-group error variances. As described in detail in Chapter 4, equality of error variances across moderator-based subgroups is an important assumption of the MMR model.

Limitations

The program MMRPOWER is the most complete tool designed so far to compute the power of an MMR test. As described in this chapter, it con-

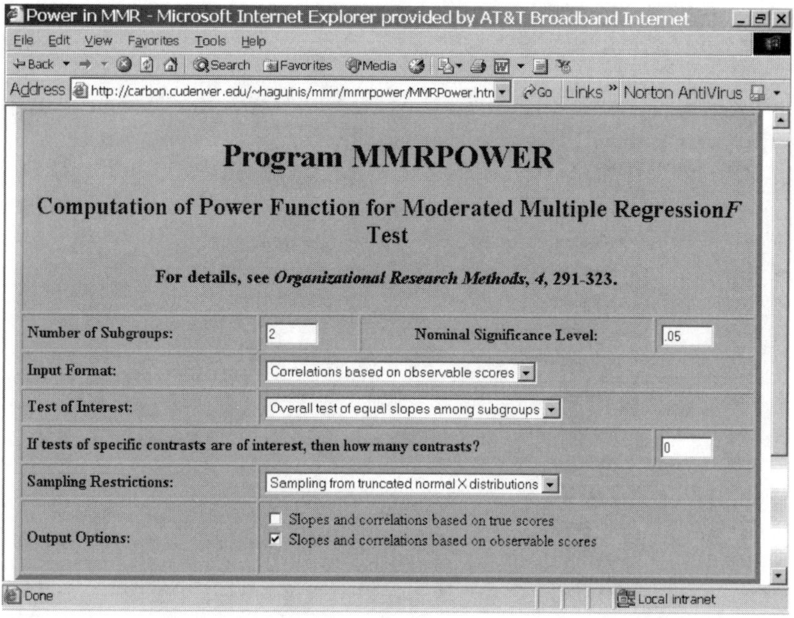

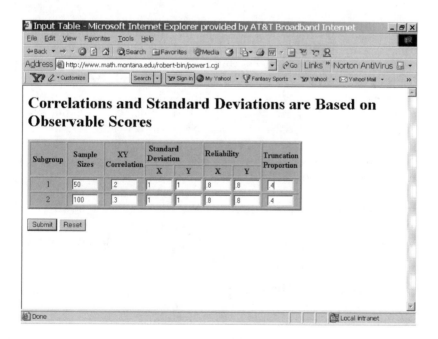

FIGURE 7.7. Illustrative initial input screen for the program MMRPOWER.

FIGURE 7.8. Illustrative second input screen for the program MMRPOWER.

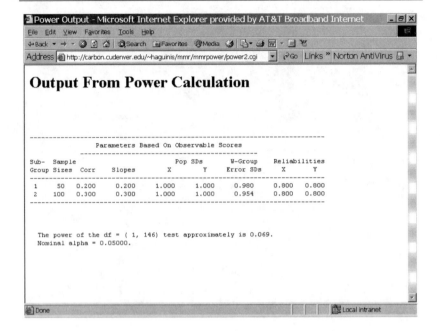

FIGURE 7.9. Illustrative output screen for the program MMRPOWER.

siders all the factors known thus far to have an impact on power. However, the program has two potential limitations. The limitations are not
really due to the program per se, but to the type and quality of information that is used as input. First, information regarding truncation and
the variance multiplying factor (when the sampling mechanism is other
than truncation) may not be available. If this information is not available, MMR users can use an estimate based on relevant literature. If a
literature-based estimate is not available, one can input a "best-case scenario" and a "worst-case scenario." The best-case scenario includes a
truncation proportion of 0.00 and assumes no truncation or a variance
multiplying factor of 1.00 and assumes no variance difference between
the sample and the population, whereas the worst-case scenario involves a severe truncation proportion such as 0.75 or a severe variance
multiplying factor of 0.25. Given the absence of information on truncation and the variance multiplying factor, MMR users will know that the
power of their MMR test lies somewhere between the best-case and
worst-case situations. The second limitation is that if the true score options are chosen, sample-based statistics must be used to estimate parameters. However, if these sample-based statistics are not good esti-

mates of the true parameters, the program may not yield an accurate power value. However, the program will yield an accurate power value as long as the parameter estimates are accurate.

RELATIVE IMPACT OF THE FACTORS AFFECTING POWER

As noted in the first section of this chapter, the computation of power is useful for at least six reasons. An additional use of the computer programs described in this chapter is to assess the relative impact of the various factors that affect power. For example, one could input what is considered adequate values for sample size and moderating-effect magnitude, and then vary measurement error on X to assess the effects of this manipulation on power. Alternatively, one could hold measurement error at a constant value and vary truncation for one or more subgroups. In this way, one could gain a better understanding of the relative impact of each of the factors on power.

Aguinis, Boik, and Pierce (2001) used the program MMRPOWER with a large number of conditions and concluded that the following is a rank ordering of factors that affect the power of MMR in order of importance:

1. Moderating effect magnitude
2. Total sample size
3. Sampling restriction on X
4. Measurement error.

Some authors have argued that measurement error is perhaps the most important culprit for MMR's lower power (Ganzach, 1998; Kromrey & Foster-Johnson, 1999). However, this assertion has not been based on a research study including several factors known to affect power. Moreover, the empirical assessment by Aguinis, Boik, and Pierce (2001) has not substantiated this claim. On the other hand, the results reported by Aguinis, Boik, and Pierce (2001) are based on specifically chosen parameter values, so they may not necessarily generalize to every single research situation. The computer programs described in this chapter can be used to conduct additional studies on the relative impact of the various factors known to affect power.

Investigators are advised to compute power given specific parameter values encountered in their research. In addition, Table 7.1 shows the statistical power for an alpha of .05 and various combinations of values for an MMR test that includes a binary moderator. In order to simplify the power calculations, which were performed using MMRPOWER, Table 7.1 makes the following assumptions: (1) Variances for X and Y are 1 in each of the groups (i.e., correlation coefficients equal regression coefficients); (2) sample size is equal across the groups; (3) there are no sampling restrictions on X; and (4) reliability is .80 for X and Y in each of the groups.

Salgado (1998) reported a median sample size of 113 for criterion-related validity studies in the personnel selection domain pub-

TABLE 7.1. Power Values (α = .05) for Various Combinations of Within-Group Slopes and Total Sample Sizes

$b_{Y.X(1)}$ $b_{Y.X(2)}$		40	80	120	160	200	240	280	320	400	500
.0	.1	.06	.07	.08	.10	.11	.12	.13	.14	.17	.20
.0	.2	.09	.14	.19	.24	.39	.34	.39	.43	.52	.61
.0	.3	.14	.26	.37	.47	.57	.65	.72	.77	.86	.93
.0	.4	.22	.43	.60	.73	.82	.89	.93	.96	.99	1.00
.0	.5	.33	.62	.81	.91	.96	.98	.99	1.00	1.00	1.00
.0	.6	.47	.80	1.00	1.00	1.00	1.00	1.00	1.00	1.00	1.00
.1	.2	.06	.07	.08	.10	.11	.12	.13	.15	.17	.20
.1	.3	.09	.14	.19	.24	.30	.35	.40	.44	.53	.62
.1	.4	.15	.27	.38	.49	.59	.67	.74	.79	.87	.94
.1	.5	.23	.45	.62	.75	.84	.90	.94	.97	.99	1.00
.1	.6	.35	.65	.84	.93	.97	.99	1.00	1.00	1.00	1.00
.2	.3	.06	.07	.09	.10	.11	.12	.14	.15	.18	.21
.2	.4	.09	.15	.20	.26	.31	.36	.41	.46	.55	.65
.2	.5	.15	.28	.41	.52	.61	.70	.76	.82	.90	.95
.2	.6	.25	.48	.66	.79	.87	.99	.96	.97	.99	1.00
.3	.4	.06	.07	.09	.10	.12	.13	.14	.16	.19	.22
.3	.5	.10	.15	.21	.27	.33	.39	.44	.49	.59	.68
.3	.6	.16	.31	.44	.56	.66	.74	.80	.85	.92	.97
.4	.5	.06	.08	.09	.11	.12	.14	.15	.17	.20	.24
.4	.6	.10	.17	.23	.30	.36	.43	.48	.54	.63	.73
.5	.6	.06	.08	.10	.12	.13	.15	.17	.19	.22	.27

Note. $b_{Y.X(1)}$ and $b_{Y.X(2)}$ are the Y on X slope for moderator-based Groups 1 and 2, respectively, and N = total sample size. Variances for X and Y are 1 in each of the groups, sample size is equal across the groups, there are no sampling restrictions on X, and reliability is .80 for X and Y in each of the groups. Cells with power $\geq$ .80 are shaded.

lished between 1983 and 1994 in applied psychology journals. Similarly, Russell et al. (1994) reported a median sample size of 103 for all validation studies of personnel selection systems published between 1964 and 1992 in *Journal of Applied Psychology* and *Personnel Psychology*. On a more encouraging note, Jaccard and Wan (1995) reviewed *American Psychological Association* journals and reported that the median sample size is 175. Table 7.1 shows that for what seems to be a fairly typical sample size of 120 in several research domains, a difference of at least .5 in regression coefficients across groups is needed for MMR to achieve a power of .80 or higher. Considering a difference between regression coefficients of .3, Table 7.1 indicates that sample size needs to be at least 300 for MMR to reach a power of .80 or higher.

CONCLUSIONS

■ This chapter described several reasons why it is important to compute power. First, it is important to compute power before a study is conducted because null findings regarding the moderating effect are inconclusive when power is low. Second, conducting a power analysis helps researchers make decisions regarding various design alternatives vis-à-vis the resources available. For example, researchers can assess whether a specific increase in sample size leads to an increase in power sufficiently substantive to justify the time and cost associated with the additional data collection effort. Third, conducting a retrospective power analysis of previously published MMR studies is useful because it may lead to the conclusion that null findings based on low-power MMR tests should be revisited. Fourth, it is also helpful to conduct a retrospective power analysis after a study has been conducted. Ideally, it is desirable to conduct the power analysis a priori. However, it is important to also conduct a power analysis once the study is completed, particularly in situations where results suggest no moderating effect. However, it is not useful to use the observed effect size as the targeted effect size. Instead, a critical effect size should be selected based on the specific research and outcome context. Finally, given the increased interest in null hypothesis significance testing and power issues, numerous journal editors and reviewers, as well as funding agencies, now require a power analysis routinely. Thus, it will become increasingly difficult to publish empirical research, particularly in the case of null findings, and request research funds based on MMR in the absence of a power analysis.

■ This chapter also described computer programs available at *www.cudenver.edu/~haguinis/mmr* to calculate the statistical power of MMR given specific research conditions. Two of these programs, POWER and MMRPWR, are based on previously published empirical results, and the third one, MMRPOWER, is based on a theory-based approximation to power. These computer programs allow researchers to compute power in planning their research design, to assess the relative impact on power resulting from varying the factors that affect power, to conduct a cost–benefit analysis of the relative advantages and disadvantages of manipulating the factors that affect power in an actual study, and to assess the power of MMR analyses reported in published studies. Given the ease of use and availability of these programs, it is no longer justified to conduct an MMR analysis without a statistical power assessment. Furthermore, it is no longer justified to conclude that the null hypothesis of no moderating effect is correct unless a power assessment has resulted in a value sufficiently high to detect a hypothesized moderating effect.

■ This chapter also described the following rank order of factors that affect the power of MMR: (1) moderating-effect magnitude, (2) total sample size, (3) sampling restriction on X, and (4) measurement error. This information is useful because it points to the factors that researchers should give priority in their attempts to maximize statistical power. Finally, the chapter included a table showing power for various combinations of total sample size and within-group regression coefficients for the case of a binary moderator. Although it is advisable to compute power given each specific research situation, this table provides a "quick and dirty" way of ascertaining sample sizes needed to detect moderating effects of various sizes. This power table also indicates that typical sample sizes used in published research in the social sciences are not sufficiently large to detect moderating effects.

■ Chapter 8 addresses MMR models that are more complex in nature than the more typical MMR model studies thus far. Specifically, Chapter 8 describes how to (1) set up four different coding schemes to analyze data sets that include moderator variables with more than two levels, (2) analyze and interpret moderating effects vis-à-vis nonlinear effects (e.g., quadratic, cubic), and (3) test and interpret MMR models including three-way and higher-order interaction effects.

8

Complex MMR Models

I ... must continue to strive for more knowledge and more power,
though the new knowledge always contradicts the old and the
new power is the destruction of the fools who misuse it.
—GEORGE BERNARD SHAW

The novelty we want is always close to the familiar.
—MASON COOLEY

This chapter addresses MMR models that are more complex in nature than the model studied so far. Many MMR users are likely to face some of the issues discussed in this chapter in their search for moderating effects.

This chapter addresses three issues regarding more complex MMR models. First, the chapter includes a discussion of how to test hypotheses regarding moderator variables with more than two levels. The discussion includes a detailed analysis and illustration of various coding schemes, their relative appropriateness for various research situations, and a demonstration of how to compute statistical power for the specific comparisons under consideration. Second, the chapter includes a discussion of the analysis and interpretation of interaction effects vis-à-vis nonlinear effects (e.g., quadratic, cubic). This section addresses the question of whether quadratic terms should be included in the MMR model routinely and offers guidelines for practice based on theory as well as Monte Carlo simulation results. Third, the chapter includes a discussion of how to test and interpret MMR models including three-way and higher-order interaction effects. This section also includes recommendations for situations when researchers may wish to test for "targeted" lower-order interactions (e.g., a specific two-way interaction in an MMR model including three predictors) without necessarily testing for all possible interactions (i.e., all three two-way interactions).

MMR ANALYSES INCLUDING A MODERATOR VARIABLE WITH MORE THAN TWO LEVELS

So far, the book has addressed the typical case where the moderator Z includes two categories (e.g., gender: male–female; ethnicity: majority–minority). In this simple and most typical case, the MMR model includes just one variable Z for which members of one category are assigned one number (e.g., 0), and members of the second category another (e.g., 1). Precisely, this is the 0–1 dummy coding scheme chosen for the illustration in Chapter 3. When the moderator includes two categories only, dummy coding is the suggested strategy because it is easy to implement and the interpretation of the results is relatively straightforward.

Although this volume has used the label "moderator" or "predictor" to refer to Z for the case involving two categories, it could also have referred to Z as a "code variable." When the moderator variable includes more than two categories, it is necessary to include more than one code variable in the MMR model. In general, the number of code variables needed is $k - 1$, where k is the number of categories or subgroups. Thus, in the simplest case including two groups, the number of code variables needed is just one because $k - 1 = 2 - 1 = 1$. However, let us consider a situation where the moderator is ethnicity, with the following three categories (i.e., $k = 3$): (1) Whites, (2) African Americans, and (3) Latinos. In this case, the MMR model needs to include two code variables (i.e., Z_1 and Z_2). Moreover, the MMR model must include the two code variables as well as the product terms between each of the code variables and the quantitative predictor X to fully represent the first-order and moderating effects. In symbols, the MMR model is the following:

$$Y = a + b_1 X + b_2 Z_1 + b_3 Z_2 + b_4 X \cdot Z_1 + b_5 X \cdot Z_2 + e \qquad (8.1)$$

where Z_1 and Z_2, taken together, represent the first-order effect of the predictor Z hypothesized to be the moderator, and $X \cdot Z_1$ and $X \cdot Z_2$ form a set that represents the interaction between X and Z (i.e., moderating effect of Z; West, Aiken, & Krull, 1996). As was the case with categorical moderators including two categories, all lower-order effects must be in the equation in addition to the product terms. Also, all code variables (i.e., $k - 1$) must be included in the equation for the product term regression coefficients to be interpretable (Serlin & Levin, 1985).

How do we choose values for Z_1 and Z_2 for each of the members of our three groups? First, we need to follow the general rule that values must be (1) different for members of different groups and (2) identical for all members of the same group (Keppel & Zedeck, 1989, Chapter 9). Next is a discussion of four types of coding schemes: dummy coding, unweighted effect coding, weighted effect coding, and contrast coding. Regardless of the chosen coding scheme, however, one typically wishes to examine whether the addition of $X \cdot Z_1$ and $X \cdot Z_2$, entered as a set in Equation 8.1, yields a statistically significant increment in R^2 above and beyond the lower-order terms (i.e., X, Z_1, and Z_2). It should be emphasized that the resulting ΔR^2, F statistic, and statistical significance level are identical regardless of the coding scheme chosen. In fact, the values for R^2, F, and p are also identical across coding schemes for Model 1 (i.e., including the intercept and first-order effects only). Thus, the test regarding the presence of a moderating effect is identical regardless of coding scheme. Next is a description of the four types of coding together with their relative appropriateness and usefulness for various research situations.

Dummy Coding

As described for the case of binary moderators in Chapter 3, in dummy coding members in one of the categories are assigned the value of 0 for each of the code variables. This type of coding is most useful when a researcher is interested in making a comparison between one target (i.e., comparison) group and the other groups. The selection of the comparison group that receives a 0 is admittedly arbitrary, but there is typically an interest in using this group as a comparison, for example, for policy-related reasons. In addition, the comparison group should be clearly defined and not be a wastebasket "other" category because subsequent comparisons with other groups may not be sufficiently meaningful (West et al., 1996).

In our example, let us choose to use the category "White" as the comparison group. Table 8.1 (the Whites as Comparison Group columns) shows that the category Whites has been assigned a 0 for each of the code variables Z_1 and Z_2, whereas the other two groups have been assigned a 1 on the code variable that will contrast each of the groups with the comparison group and a value of 0 otherwise.

Alternatively, we may be interested in, for example, using the

TABLE 8.1. An Illustration of Coding Schemes for a Categorical Moderator
Variable Including Three Categories

	Whites as comparison group		Latinos as comparison group	
	Z_1	Z_2	Z_1	Z_2
Dummy coding				
Whites	0	0	1	0
African Americans	1	0	0	1
Latinos	0	1	0	0
Unweighted effect coding				
Whites	−1	−1		
African Americans	1	0		
Latinos	0	1		
Weighted effect coding				
Whites	$-n_{AA}/n_W$	$-n_L/n_W$		
African Americans	1	0		
Latinos	0	1		
Contrast coding				
Whites	−2/3	0		
African Americans	1/3	1/2		
Latinos	1/3	−1/2		

Latino category as the comparison group. In this case, members of the
Latino group are assigned a 0 for each of the code variables. This is also
shown in Table 8.1 (i.e., the Latinos as Comparison Group columns).

Considering the Whites as Comparison Group columns in Table
8.1, the code variable Z_1 compares Whites (assigned a 0) with African
Americans (assigned a 1), whereas Z_2 compares Whites (assigned a 0)
with Latinos (assigned a 1). Assume that X has been centered prior to
the analysis. In Equation 8.1, the intercept is the mean score on Y for
members of the White group at the overall mean (across all individuals
in the three groups) of X, taking into consideration the effect of X on Y
for the White group. In addition, b_1 is the slope of Y on X for Whites
(i.e., the comparison group). That is, the regression coefficient b_1 is
identical to the coefficient for X resulting from examining the scores for
the White group only obtained through the following simpler bivariate
regression equation:

$$Y = a + bX + e \qquad (8.2)$$

The coefficients b_2 and b_3 in Equation 8.1 represent mean differences across groups regarding Y. Specifically, they are the differences in Y scores between Whites–African Americans and Whites–Latinos, respectively. Thus, for example, $b_2 = 2$ means that, on average, Y scores are 2 points higher for members of the White group as compared to members of the African American group. The coefficients b_4 and b_5 represent slope differences across groups. Specifically, they are the differences between the Y on X slope between Whites–African Americans and Whites–Latinos, respectively. So, for example, $b_5 = .5$ means that the slope for members of the White group is .5 points steeper than the slope for members of the Latino group. In short, the coefficients for the first-order effects of the code variables provide information regarding mean differences regarding Y across groups, whereas the coefficients for the product terms provide information regarding Y on X slope differences across groups (Cohen et al., 2003, Chapter 9).

Finally, as noted earlier, the interpretation of the coefficients in Equation 8.1 assumes that the quantitative predictor X has been centered prior to the analysis. As described in detail in Chapter 3, in some cases a 0 value for X may be difficult to interpret, or even not meaningful (i.e., zero may fall outside of the X scale). If X has not been centered prior to obtaining the regression estimates, the intercept in Equation 8.1 would be interpreted as the mean score on Y for members of the White group for whom $X = 0$. Similarly, if X is not centered, the coefficients b_2 and b_3 would be the differences in Y scores between Whites–African Americans and Whites–Latinos for individuals for whom $X = 0$. However, the interpretation of the coefficients b_4 and b_5, as well as the R^2 and significance level associated with Model 1 (i.e., first-order effects only) and Model 2 (first-order and interaction effects), would remain unchanged.

Unweighted Effect Coding

For unweighted effect coding, also called effects coding, members in one of the categories (i.e., the "focal" group) are assigned the value of –1 for each of the code variables. Members of the other two groups are given a +1 on one of the code variables and a 0 on all the other (just one other in our illustration involving $k = 3$ and two code variables). Table 8.1 shows the resulting values for each of the two code variables using the group *Whites* as an illustrative focal group.

In contrast to dummy coding, the regression coefficients do not provide information about the comparison between the focal group assigned the value of −1 on each of the code variables and each of the other groups. Instead, comparisons are made in reference to the mean (of Y scores, or of the slope of Y on X) across all three groups. So, assuming X has been centered, the intercept in Equation 8.1 is the overall mean for Y across the three groups (holding X constant) computed as follows:

$$\frac{M_W + M_{AA} + M_L}{3} \tag{8.3}$$

where M_W, M_{AA}, and M_L are the mean Y scores for the White, African American, and Latino groups, respectively. The coefficient b_1 in Equation 8.1 is the mean slope for the regression of Y on X across the groups computed as follows:

$$\frac{b_{Y.X(W)} + b_{Y.X(AA)} + b_{Y.X(L)}}{3} \tag{8.4}$$

where $b_{Y.X(W)}$, $b_{Y.X(AA)}$, and $b_{Y.X(L)}$ are the slopes of Y on X for the White, African American, and Latino groups, respectively.

The coefficient b_2 in Equation 8.1 is the difference between the mean Y scores for the African American group and the mean Y scores across the three groups (holding X constant), and the coefficient b_3 is the difference between the mean Y scores for the Latino group and the mean Y scores across the three groups (also holding X constant).

The coefficients b_4 and b_5 are the differences between the Y on X slope between African Americans and the mean slope, and Latinos and the mean slope, respectively. Note that the difference between the White group and the mean slope across groups can be computed as follows: −$(b_1 + b_2)$.

Weighted Effect Coding

Weighted effect coding is similar to effect coding. However, it differs from effect coding in that weighted effect coding takes into account the sample sizes in each of the moderator-based categories. This type of coding is useful when sample sizes differ substantially across groups and one may wish to take into account sample size in each of the groups in computing the average Y scores and average Y on X slopes. When sample sizes across the groups are identical, the coding scheme as well as results and interpretation of unweighted and weighted effect coding are identical. However, results become increasingly different as sample sizes across groups become more divergent.

In assigning values to members of the various groups, Table 8.1 shows that members of the focal group (i.e., White for the sake of this illustration) are assigned a value based on the sample size of the focal group in the denominator and the sample size of the group assigned a value of 1 for each code variable in the numerator. Thus, for Z_1 the value is $-n_{AA}/n_W$ and for Z_2 the value is $-n_L/n_W$.

When unweighted coding is used, the intercept in Equation 8.1 is the overall *weighted* mean for Y across the three groups (holding X constant) computed as follows:

$$\frac{n_W M_W + n_{AA} M_{AA} + n_L M_L}{n_W + n_{AA} + n_L} \tag{8.5}$$

Similarly, the coefficient b_1 is the mean *weighted* slope for the regression of Y on X across groups (holding X constant) computed as follows:

$$\frac{n_W b_{Y.X(W)} + n_{AA} b_{Y.X(AA)} + n_L b_{Y.X(L)}}{n_W + n_{AA} + n_L} \tag{8.6}$$

The remaining regression coefficients are interpreted in the same way as for the unweighted effect codes, but comparisons are made in reference to the weighted mean Y (i.e., Equation 8.5 instead of Equation 8.3) and to the weighted mean Y on X slope (i.e., Equation 8.6 instead of Equation 8.4). The difference between the Y on X slope for the White group and the overall weighted Y on X mean slope can be computed as follows: $-[(-n_{AA}/n_W)b_1 + (-n_L/n_W)b_2]$.

Contrast Coding

Contrast coding is used when a researcher has a specific a priori interest in performing targeted comparisons between certain Y on X slopes (e.g., Whites vs. African Americans and Latinos combined, African Americans vs. Latinos) (Rosenthal & Rosnow, 1985; Rosenthal, Rosnow, & Rubin, 2000). There are three rules that need to be followed when creating code variables following contrast coding (West et al., 1996):

1. The sum of the weights for each code variable must equal 0.
2. The difference between the value of the positive weights and negative weights should equal 1 for each of the code variables.
3. The sum of the products of each pair of codes should equal 0.

Let us confirm that the contrast coding values shown in Table 8.1 comply with the three above rules. Regarding rule 1, for Z_1: $-2/3 + 1/3 + 1/3 = 0$, and for Z_2: $0 + 1/2 + (-1/2) = 0$. Regarding rule 2, for Z_1: $1/3 - (-2/3) = 1$, and for Z_2: $1/2 - (-1/2) = 1$. Regarding rule 3, the product of the values for Whites is $(-2/3) \cdot 0 = 0$; for African Americans,

$(1/3) \cdot (1/2) = 1/6$; and for Latinos, $(1/3) \cdot (-1/2) = -1/6$, resulting in a sum of the products of $0 + 1/6 - 1/6 = 0$. Thus, the code variables shown in Table 8.1 do comply with the three contrast coding rules.

The resulting intercept and coefficient b_1 in Equation 8.1 are identical to those obtained using unweighted effect coding. That is, the intercept is the overall mean for Y across the three groups, and b_1 is the mean slope for the regression of Y on X across groups (holding X constant). The coefficient b_2 is the difference between mean Y scores for the group of Whites versus the other two groups combined, and the coefficient b_3 is the difference between mean Y scores for the African American versus the Latino group. Similarly, the coefficient b_4 is the difference between the Y on X slope for the group of Whites versus the combined Y on X slope for the other two groups, and the coefficient b_5 is the difference between the Y on X slope for the African American versus the Latino group.

How to Choose a Coding Scheme: The Importance of Theory

Typically, in testing whether the categorical moderator has an effect on the relationship between X and Y, researchers want to first answer the overall question of whether there is a moderating effect at all. This can be achieved by examining the size and statistical significance for ΔR^2. However, similar to ANOVA (Rosenthal & Rosnow, 1985) and meta-analysis contexts (Aguinis & Pierce, 1998c), a researcher may have specific hypotheses regarding the form of the interaction effect and, therefore, may choose to examine b_4 or b_5, but not both. Alternatively, a researcher may wish to examine b_4 and/or b_5 as a follow-up to a finding that the set of $X \cdot Z_1$ and $X \cdot Z_2$ predictors increased the proportion of Y variance explained. In these situations one should be aware that, based on the preceding discussion, the chosen coding scheme has profound implications for the meaning of the regression coefficients. Because each coding scheme represents a different way to partition the total variance associated with the interaction, the same regression coefficient (e.g., b_4) answers a different question for each coding scheme and, therefore, its value and statistical significance level also vary (Schoorman et al., 1991).

The key factor involved in choosing which coding scheme to use is the underlying theory. For example, if the goal of the study is to compare the slope for one group (e.g., control or baseline group) with the slope of each of the other groups, then dummy coding should be the preferred strategy. On the other hand, if one wishes to compare the slope of one group vis-à-vis the mean slope for all groups included in the study, effect coding is most appropriate. Last, if one wishes to make specific comparisons involving one group vis-à-vis the combination of other groups, then contrast coding will perform this task best. In short, the choice for a coding scheme must be guided by the underlying hypotheses regarding the form of the interaction effect (Irwin & McClelland, 2001). It should be clear, however, that the coding system is irrelevant in the sense that it is the same equation that is estimated and that the test of the moderating effect yields identical results regardless of coding scheme. However, the coding system used determines the interpretation of the form of the moderating effect.

One final consideration is that if one wishes to examine specific contrasts without necessarily examining the overall interaction effect first, this test should be conducted before examining the data and should be supported by a good conceptual framework (Bobko, 1986; Boik, 1979, 1993). Engaging in "data snooping" can lead to a Type I error and the incorrect conclusion that there is a moderating effect (Bobko & Russell, 1994; Rasmussen, 1989).

Computing Statistical Power

Aguinis, Boik, and Pierce's (2001) program MMRPOWER, available at *www.cudenver.edu/~haguinis/mmr*, allows for the computation of power for MMR analyses including up to 20 categories for the moderator variable. Therefore, the description and illustration provided in Chapter 7 regarding a binary moderator generalizes to moderators including more than two levels. However, in situations involving moderators with more than two categories, one needs to specify that power computations be obtained for specific contrasts. Figure 8.1 shows the opening screen for MMRPOWER, where one needs to choose the option "Test of specific contrasts of slopes" and specify the number of desired contrasts. Let us continue with the preceding example regarding ethnicity and use the contrast coding scheme shown in Table 8.1. Let us further assume that sample size is 100 in each of the three groups, $SD = 1$ for X and Y across groups so that correlations equal slopes, and X–Y correlations are .20, .45, and .60, respectively, for each of the three groups. Figure 8.2 shows the resulting MMRPOWER screen.

Figure 8.3 shows the resulting output, including information about the power for the overall interaction test (i.e., .80) as well as the power for each of the two contrasts. Specifically, Figure 8.3 shows that the hypothesis test that the slope of Y on X differs for the White versus the African American and Latino groups combined is conducted at a power of .82, whereas the hypothesis test that the slope is different for the African American versus the Latino group is conducted at a power of .20. Ideally, this power analysis is conducted before data collection and would

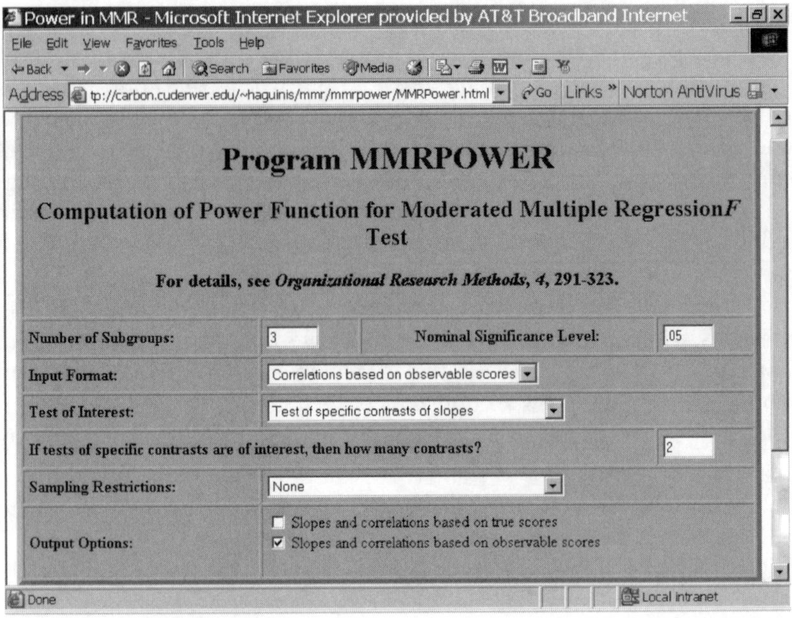

FIGURE 8.1. Initial screen for program MMRPOWER.

therefore lead to the conclusion that larger sample sizes are needed for the difference between the African American and Latino groups to be detected. On the other hand, if this power analysis is conducted in a retrospective fashion (and, assuming the targeted effect size is of the same magnitude as the observed effect size), a conclusion of no difference in the slopes between the African American and Latino groups should be taken with caution. Next, the chapter addresses a second type of more complex MMR models: Models including nonlinear terms.

LINEAR INTERACTIONS AND NONLINEAR EFFECTS: FRIENDS OR FOES?

In some areas in the social sciences, researchers hypothesize, and wish to test for, nonlinear relationships between a criterion and a predictor (e.g., Avolio, Waldman, & McDaniel, 1990). For example, Lubinski and Humphreys (1990) tested whether the relationship between the criterion *advanced mathematics skills* and the predictor *mathematical aptitude* was curvilinear, as shown in Figure 8.4. Specifically, they tested this hypothesized relationship using data on 400,000 participants in the Project Talent Data Bank (Flanagan et al., 1962).

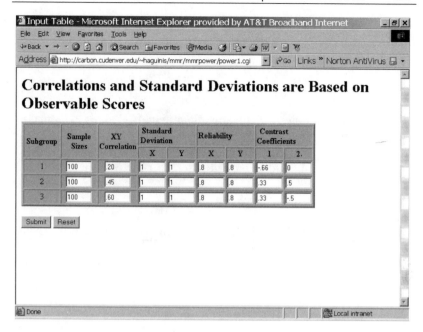

FIGURE 8.2. Contrast coefficient screen for program MMRPOWER.

Output From Power Calculation

Subgroup	Sample sizes	Corr	Slopes	Pop SDs		W-Group Error SDs	Reliabilities	
				X	Y		X	Y
1	100	0.200	0.200	1.000	1.000	0.980	0.800	0.800
2	100	0.450	0.450	1.000	1.000	0.893	0.800	0.800
3	100	0.600	0.600	1.000	1.000	0.800	0.800	0.800

Parameters Based On Observable Scores

The power of the df = (2, 294) test approximately is 0.803.
Nominal alpha = 0.05000.

Contrast	Power
1	0.819
2	0.202

FIGURE 8.3. Output screen for contrast analysis for program MMRPOWER.

The hypothesized effect shown in Figure 8.4 can be tested using multiple regression with the following equation (Cohen & Cohen, 1983):

$$Y = a + b_1 X + b_2 X^2 + e \qquad (8.7)$$

where, in this example, Y is advanced mathematical skills, X is mathematical aptitude, and the regression coefficient b_2 gives us information on whether there is a curvilinear relationship between X and Y. Also, one could include an additional cubic term in the equation:

$$Y = a + b_1 X + b_2 X^2 + b_3 X^3 + e \qquad (8.8)$$

where b_3 would provide information regarding whether the relationship between X and Y is U-shaped.

Note that Equation 8.7 can be rewritten in the more familiar MMR way as follows:

$$Y = a + b_1 X + b_2 X \cdot X + e \qquad (8.9)$$

where the relationship between X and Y is moderated by X itself. Indeed, a perusal of Figure 8.4 shows that the steepness of the slope regressing Y on X varies at various levels of X such that the relationship is

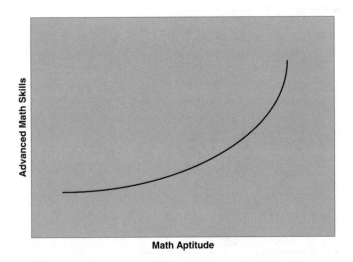

FIGURE 8.4. Hypothesized curvilinear relationship between mathematical aptitude and advanced mathematics skills.

near zero for low scores on mathematics aptitude and strong and positive for medium and high scores.

In short, in addition to testing linear interaction effects, researchers may have theory-based reasons to test for the presence of nonlinear (e.g., quadratic, cubic, quartic, and higher-order) effects. In fact, it is possible that both a linear interaction *and* a nonlinear effect may be hypothesized in the context of the same study. For instance, the Project Talent Data Bank includes data for both female and male high school students, and there was a relevant theory-related reason to test (1) whether mathematical aptitude predicted advanced mathematics skills differentially for boys and girls (i.e., moderating-effect hypothesis), as well as (2) whether the relationship between mathematical aptitude and advanced mathematics skills was curvilinear, as is shown in Figure 8.4 (i.e., quadratic effect hypothesis) (Lubinski & Humphreys, 1990).

A potential problem regarding testing interaction and nonlinear effects is that variance in Y due to the quadratic effect of X (represented by the term X^2) may be unduly attributed to the X by Z interaction (represented by the term $X \cdot Z$). To illustrate this phenomenon, Lubinski and Humphreys (1990) tested the first-order and interactive effects of X (i.e., math aptitude) and Z (gender) by sequentially forcing X, Z, and $X \cdot Z$ in the regression equation predicting Y (i.e., the traditional MMR model). Results showed that, by entering the product term, R^2 was increased significantly and thus there was evidence for an interaction effect between X and Z. However, in a second set of analyses using the same data, Lubinski and Humphreys first forced the first-order effects of X and Z, and then simultaneously entered the terms carrying information about the linear interaction between X and Z (i.e., $X \cdot Z$) as well as the terms carrying information about the quadratic effects of X and Z (i.e., X^2 and Z^2) resulting in the following equation:

$$Y = a + b_1 X + b_2 Z + b_3 X \cdot Z + b_4 X^2 + b_5 Z^2 + e \qquad (8.10)$$

Results regarding Equation 8.10 showed that virtually all the Y variance that had been attributed to $X \cdot Z$ in the first analysis was completely absorbed by the X^2 term. Given this result, Lubinski and Humphreys (1990) concluded that the moderating effect found in the first analysis was spurious because a curvilinear trend explained the relationship between X and Y better than the X by Z linear interaction.

In more general terms, further simulation work has demonstrated that the failure to include quadratic terms in the MMR model can lead to the discovery of a "false" interaction as well as to failing to discover an existing interaction effect (Ganzach, 1997, 1998). Thus, based on these results, Lubinski and Humphreys recommended that squared terms be routinely included as part of any MMR model, and others have gone as far as to recommend that squared terms be entered hierarchically before the product term (Cortina, 1993) (i.e., the order of entry should be X and Z at Step 1, X^2 and Z_2 at Step 2, and $X \cdot Z$ as the last step).

Reacting to the recommendation that nonlinear terms be routinely included in MMR analyses, Shepperd (1991) argued that the systematic testing of quadratic effects in MMR "may be unwise and, at the very least, should be viewed cautiously" (p. 316). The reason for this statement is that when the predictor X and the moderator Z are highly correlated (i.e., $r_{XZ} \to 1.00$), $X \cdot Z$ and X^2 share substantial variance. However, as r_{XZ} decreases, the overlap in variance between $X \cdot Z$ and X^2 is reduced, and so is the probability of finding a spurious moderator. Subsequently, MacCallum and Mar (1995) provided empirical confirmation for this conceptual analysis. Consequently, in situations when the correlation between the predictor X and the moderator Z is small, systematically including X^2 and Z^2 in MMR analyses without a priori predictions about these effects may lead to spurious findings (i.e., falsely rejecting a null hypothesis) regarding quadratic effects.

MacCallum and Mar (1995) found that systematically including quadratic terms in the MMR model can lead to a correct assessment of the presence of interaction and quadratic effects (unless sample size, reliability, and effect size are low) when the population model is interactive (and not quadratic). However, systematically using Equation 8.10 when the population model is quadratic is likely to lead to incorrect conclusions about the presence/absence of interaction and/or quadratic effects under a variety of conditions. Of course, we do not know the true nature of the population model, and this is precisely why we are implementing inferential statistical procedures in the first place. Therefore, the only way to know whether, from a data analysis perspective, it may be useful to include quadratic terms in the MMR model is to adhere to the following three guidelines derived from MacCallum and Mar's simulation results:

- The correlation between X and Z is $\geq .50$.
- Reliabilities for X and Z are $\leq .70$.
- Sample size is approximately 75 or greater (Kromrey & Foster-Johnson, 1999, argued that MacCallum and Mar's simulation included moderating-effect sizes unrealistically large and, thus, the guideline regarding sample size should be revised to approximately 175 or greater).

Should researchers include quadratic terms routinely in their MMR analysis (Ganzach, 1998, Lubinski & Humphreys, 1990)? Moreover, should quadratic terms be included hierarchically *before* the product term in the MMR model (Cortina, 1993)? Simulation work suggests that, from a data analysis point of view, these practices are warranted only when predictors are highly correlated and not highly reliable. However, in addition to this data-based consideration, a more important decision-making criterion is what the theory is underlying the proposed quadratic and/or interaction model. Does a researcher have a specific hypothesis about the presence of a quadratic effect, or does the proposed model include an interaction effect only? Routinely including quadratic terms in the MMR model in the absence of specific hypotheses is consistent with exploratory as opposed to confirmatory research (Kromrey & Foster-Johnson, 1999; Tukey, 1977). Results based on exploratory research, which should be recognized as such, should be taken with great caution. Such results must be replicated using independent samples, particularly if the model includes unexpected and complex nonlinear effects. As noted in other sections of this volume, there is no statistical technique that can substitute for good thinking. In short, "inspection of quadratic and other nonlinear terms, unless specifically predicted, should be viewed as exploratory and treated prudently" (Shepperd, 1991, p. 316). Furthermore, "perhaps the best answer is not in better statistics but in better thinking about the nature of inquiry" (Kromrey & Foster-Johnson 1999, p. 412).

Therefore, the recommendation is to include nonlinear terms in the MMR model when the following two conditions are met:

1. There is a specific theory-based rationale for including nonlinear terms.
2. Correlations between predictors is high, reliability of predictors is low, and sample size is large.

TESTING AND INTERPRETING THREE-WAY AND HIGHER-ORDER INTERACTION EFFECTS

So far this chapter has focused on the typical moderation situation including a criterion, one predictor, and one moderator. However, an MMR model can handle two or more predictors that may be hypothesized to interact with each other (e.g., Aguinis & Stone-Romero, 1997). An important requirement is that all lower-order terms be entered in the equation. This is similar to the requirement that X and Z be entered in the regression equation in the presence of the product term $X \cdot Z$.

Recall the fictitious illustration used in previous chapters where we are predicting salary increase Y based on the quantitative variable job performance score X and the moderator tenure status Z (i.e., tenured vs. untenured). Let us assume that this study includes the additional predictor gender (W), which is hypothesized to interact with job performance and tenure status in predicting salary increase. In other words, the hypothesis is that the effect of performance on salary increase is moderated by the gender by tenure interaction. Therefore, the MMR model is now the following:

$$Salary = a + b_1 Perf + b_2 Tenure + b_3 Gender + b_4 Perf \cdot Tenure + \quad (8.11)$$
$$b_5 Perf \cdot Gender + b_6 Gender \cdot Tenure +$$
$$b_7 Perf \cdot Gender \cdot Tenure + e$$

Note that because the equation includes a three-way product term, all lower-order effects are also part of the equation (i.e., all first-order and second-order effects). The three-way product term does not need to be entered in a separate step once all the lower-order terms are in the equation, but it needs to be entered simultaneously or after the lower-order terms are already entered.

The analysis and interpretation of the MMR model including a two-way interaction that was described in detail in Chapter 3 generalizes to the situation involving a three-way (or higher-order) interaction. All the considerations regarding interpretation of lower-order effects in the presence of higher-order effects as well as centering of quantitative variables are relevant.

It is possible, and usually desirable, to interpret the two-way interaction effects in the presence of a three-way interaction (see a similar discussion regarding the interpretation of first-order effects in the presence of a two-way interaction in Chapter 3). For example, assume we are using the same dummy coding for *Tenure* as in Chapter 3, where tenured = 0 and untenured = 1. The coefficient b_6 in Equation 8.11 is interpreted as the difference in the slope of *Salary* on *Gender* for the tenured minus the untenured group at the zero point for *Performance*. Recall that unless the quantitative predictor X is centered, the interpretation of lower-order effects may be virtually meaningless because the zero point may not even be defined on the scale (as in this case in which *Performance* was measured on a scale ranging from 1 to 5). This is a good example of such a situation because a score of zero is outside of the *Performance* scale. However, centering is one of several procedures that makes the zero point easier to interpret by equating it to the mean. Other procedures include making zero the midpoint or neutral point on the scale and making the zero point the median score (see Chapter 3 for details).

In terms of the interpretation of the three-way interaction, if b_7 is statistically significant (or if R^2 change after introducing the three-way product term is statistically significant), this means that the interaction effect of two predictors (e.g., *Performance* by *Tenure*) on the criterion depends on the value of the third predictor (e.g., *Gender*). In other words, the moderating effect of tenure on the performance–salary relationship is moderated by gender. Equivalently, the effect of performance on salary increase is moderated by the gender by tenure interaction. Recall that the interpretation of moderating effects is symmetrical. Thus, depending on which is the focal moderator variable in our research, we could also interpret the effect as follows: The moderating effect of gender on the performance–salary increase relationship is moderated by tenure.

As in the case of the interpretation of two-way interactions, it is generally useful to plot the three-way interaction. A similar procedure as that followed in Chapter 3 is implemented. First, we need to construct the regression equation for each of the four groups (i.e., tenured-female, untenured-female, tenured-male, and untenured-male). *Tenure* was coded as tenured = 0 and untenured = 1 and *Gender* was coded as male = 0 and female = 1. Therefore, reworking Equation 8.11 for the tenured (0) female (1) group yields the following:

$$Salary = a + b_1 Perf + b_2 Tenure + b_3 Gender + b_4 Perf \cdot Tenure + \quad (8.12)$$
$$b_5 Perf \cdot Gender + b_6 Gender \cdot Tenure +$$
$$b_7 Perf \cdot Gender \cdot Tenure + e$$

$$Salary = a + b_1 Perf + b_2(0) + b_3(1) + b_4 Perf(0) + b_5 Perf \cdot (1) +$$
$$b_6 \cdot (1) \cdot (0) + b_7 Perf \cdot (1) \cdot (0) + e$$

Tenured-female: *Predicted Salary* $= a + b_3 + Perf(b_1 + b_5)$

Reworking Equation 8.11 for the untenured (1) female (1) group yields the following:

$$Predicted\ Salary = a + b_1 Perf + b_2 + b_3 + b_4 Perf + b_5 Perf + \quad (8.13)$$
$$b_6 + b_7 Perf$$

Untenured-female: *Predicted Salary* $= a + b_2 + b_3 + b_6 +$
$Perf(b_1 + b_4 + b_5 + b_7)$

And, reworking Equation 8.11 for the tenured (0) male (0) and untenured (1) male (0) groups yields the following:

Tenured-male: *Predicted Salary* $= a + b_1 Perf$ $\qquad\qquad$ (8.14)

Untenured-male: *Predicted Salary* $a + b_2 + Perf(b_1 + b_4)$ $\qquad$ (8.15)

As described in Chapter 3, it is recommended that we choose values of 1 *SD* above and below the mean for *Performance* in Equations 8.12–8.14 (Cohen et al., 2003, Chapter 7). These values, together with the regression coefficients obtained by the resulting Equation 8.11, are used to solve Equations 8.12–8.14 and plot the slope for *Salary* on *Performance* for each of the four groups.

As an alternative plotting procedure, there are computer programs available for both SAS and SPSS that allow for plotting regression models including the following types of interactive models (O'Connor, 1998, also available on the Web at *http://flash.lakeheadu.ca/~boconno2/simple.html*):

- Interactions between two quantitative predictors.
- Interactions between a quantitative predictor and a binary moderator.
- Interactions between a quantitative predictor and a three-category moderator.
- Interactions between a quantitative predictor and a four-category moderator.
- Interactions including 3 variables (i.e., two quantitative predictors and a binary moderator).

In short, the procedure to analyze and interpret three-way and higher-order interactions are generalizations of those implemented to analyze and interpret two-way interactions. There is no noticeable difference in the analysis or interpretation.

Examining "Targeted" Lower-Order Interactions

In some situations, researchers may wish to test for the presence of a specific (i.e., "targeted") two-way interaction effect based on a model including three or more predictors. Let us return to the previous example where salary is the criterion, and the predictors are performance score, gender, and tenure. Assume there is no compelling reason to test for the presence of the three-way interaction. Thus, the MMR model is the following:

$$Predicted\ Salary = a + b_1 Perf + b_2 Tenure + b_3 Gender + \qquad (8.16)$$
$$b_4 Perf \cdot Tenure + b_5 Perf \cdot Gender + b_6 Gender \cdot Tenure$$

Now assume a researcher is not even interested in testing all three two-way interactions. Instead, he or she wishes to test for the presence of the *Tenure* by *Performance* interaction only. That is, this researcher is interested in testing H_0: β_4 only. Is it appropriate to have a Model 1 including the first-order effects of *Performance, Tenure,* and *Gender,* and Model 2 including only the *Performance* by *Tenure* product term? In other words, is the following model appropriate?

$$Predicted\ Salary = a + b_1 Perf + b_2 Tenure + b_3 Gender + \qquad (8.16)$$
$$b_4 Perf \cdot Tenure$$

This is a situation faced by researchers using not only MMR, but also other data analysis techniques. For example, in the context of a one-way ANOVA addressing the effect of four types of drugs on reaction speed of mice, one first typically assesses whether there is an overall effect for the drug. Then, as a second step, one would examine specific comparisons to better understand what is causing the overall (i.e., omnibus) effect.

The recommendation is the following. First, a researcher would test the overall effect of all two-way interactions combined by testing whether the R^2 change from the model with first-order effects to the model including all three two-way interactions is statistically significant (i.e., Equation 8.16). Then, if the omnibus test is statistically significant, as a second step a researcher would examine the regression coefficient for the two-way interaction of interest to see if this particular coefficient is statistically significant. This recommendation follows the logic that a researcher first examines the omnibus effect and then proceeds to examine specific, more in-depth, comparisons.

CONCLUSIONS

■ This chapter showed that the MMR model examined in previous chapters including a moderator with just two levels can be generalized to a situation involving a moderator with more than two levels (i.e., $k > 2$). The procedure consists of creating $k - 1$ code variables. Choosing the coding scheme is a key consideration because the choice leads to the attachment of different meanings to each of the code variables. This chapter discussed the following four types of coding schemes: dummy coding, unweighted effect coding, weighted effect coding, and contrast coding. Theory and the expected form of the interaction effect are the key factors that dictate the coding scheme we choose to implement. The choice of a coding scheme has profound implications for the meaning of the resulting regression coefficients. Because each coding scheme represents a different way to partition the total variance associated with the interaction, the same regression coefficient answers a different question for each coding scheme and, therefore, its value and statistical significance level also vary. On the other hand, the resulting ΔR^2, and associated F statistic and statistical significance level, are identical regardless of the coding scheme chosen. Thus, the test regarding the presence of a moderating effect is identical regardless of coding scheme.

- MMR users should be aware that the tools described in previous chapters regarding $k = 2$ situations also apply to $k > 2$ situations. For example, the program ALTMMR can be used to assess whether the homogeneity of error variance assumption is violated (see Chapter 4), and the program MMRPOWER can be used to compute the statistical power for the MMR test (see also Chapter 7).

- A second topic addressed by this chapter is the potential inclusion of nonlinear terms in the MMR equation. In the past few years several articles have been published regarding the potential overlap between interaction and nonlinear effects. This body of literature has led to the question of whether quadratic terms should be included routinely in every MMR model. This chapter reviewed the available empirical literature, and the conclusion is that quadratic effects should be included in MMR models when the correlations between predictors is high, reliability of predictors is low, and sample size is large. Furthermore, including quadratic terms in the MMR model without specific hypotheses is considered to be exploratory research and any findings resulting from such analysis should be replicated using independent samples. Thus, a key factor in deciding whether to include quadratic terms in an MMR model is whether there is a specific theory-based rationale for including such terms.

- This chapter also showed how the MMR model generalizes easily to situations including three or more predictors that may be hypothesized to interact with each other. All the considerations regarding interpretation of lower-order terms as well as centering discussed in previous chapters are still relevant. That is, lower-order terms need to be entered in the equation, and the interpretation of lower-order coefficients is affected by the presence of a higher-order interaction effect. Thus, unless quantitative predictors are centered, the interpretation of lower-order effects may be virtually meaningless. This section of the chapter also addressed situations wherein researchers may wish to test for the presence of a targeted two-way interaction effect. The recommendation is to follow the logic that a researcher first examines the omnibus effect and then proceeds to examine specific, more in-depth, comparisons.

■ The next chapter addresses additional topics regarding the interpretation of moderating effects. Specifically, Chapter 9 includes a discussion of (1) how to compute indicators of the "practical significance" of a moderating effect, (2) the potentially adverse consequences of using the signed coefficient rule to interpret the form of the moderating effect, and (3) the importance and consequences of not specifying clearly which variable is the criterion and which variable is the moderator in the MMR model.

9

Further Issues in the Interpretation of Moderating Effects

All human knowledge takes the form of interpretation.
—WALTER BENJAMIN

All meanings, we know, depend on the key of interpretation.
—GEORGE ELIOT

Chapters 2, 3, and 8 addressed how to interpret MMR results. This chapter addresses three additional issues in the interpretation of moderating effects. First, there is a discussion of how to assess whether the moderating effect is practically significant. This section considers several procedures that fall into one of two different types of strategies that can be implemented to assess the extent to which the moderator provides (1) improvement of model fit or (2) improvement of prediction. Second, the chapter discusses the commonly used "signed coefficient rule" for interpreting moderating effects. A case is made against the use of this practice. Finally, this chapter includes a discussion of the consequences of not specifying clearly which variable is the criterion and which variable is the moderator in the MMR model. This third section shows that the criterion and predictor variables are not interchangeable and that swapping them can lead to discrepancies in the interpretation of the nature of the moderating effect. This phenomenon is illustrated using the pay for performance data set introduced in Chapter 3.

IS THE MODERATING EFFECT PRACTICALLY SIGNIFICANT?

As noted in Chapter 5, one of the main reasons for the current controversy over null hypothesis significance testing is that some researchers misinterpret results of inferential analysis. This issue, combined with the typically low statistical power of MMR, makes it imperative that researchers compute an estimate of the population moderating effect in addition to the more traditional statistical significance test. Estimating the moderating effect size is important because it allows researchers to understand whether, in addition to statistical significance, the effect is likely to have "practical" significance. In general, there are seven ways to estimate the practical significance of the moderating effect. These measures can be classified as measures of improved fit or measures of improved prediction, and there is a subtle difference between these two types of measures. The measures of improved fit refer to the extent to which the addition of the product term to the model improves the proportion of Y variance explained, whereas the measures of improved prediction refer to whether the addition of the product term changes our understanding regarding the steepness of the slope of Y on X across the moderator-based subgroups. Each of these procedures is described next.

Measures of Improved Fit

Difference in Correlations across Moderator-Based Subgroups

The first method used to understand the strength of the moderating effect has been labeled "subgroup analysis" (e.g., Arnold, 1982). This method entails computing the correlation between X and Y for each of the moderator-based subgroups (i.e., r_1 and r_2), and then computing the difference between the correlations, usually in absolute terms (i.e., $|r_1 - r_2|$) (Aguinis & Stone-Romero, 1997). For example, assuming a moderator with two levels (men, women), a researcher would obtain the correlation between X and Y for men (e.g., $r_1 = .20$) and the correlation for women (e.g., $r_2 = .40$), resulting in an absolute difference of $|r_1 - r_2| = .20$.

One can also compute the coefficients of determination for each group (i.e., r_1^2 and r_2^2), which represent the proportion of variance in Y explained by X in each of the groups. For example, if $r_1 = .2$ and $r_2 = .4$, then $r_1^2 = (.2)(.2) = .04$ and $r_2^2 = (.4)(.4) = .16$, which indicate that the predictor X explains 4% of variance in Y for the first group and 16% of variance in Y for the second group. In addition, the difference in Y variance explained is $r_2^2 - r_1^2 = 12\%$ between the groups in favor of the second group. This method provides information regarding the relative proportion of Y variance explained by the various groups. The raw difference between coefficients of determination can be used, as opposed to the absolute difference, because coefficients of determination cannot take on negative values.

An important caveat is needed regarding the use of correlation coefficients in assessing the strength of moderating effects. The correlation coefficient and the regression coefficient are related such that $b = r(s_y/s_x)$. Thus, $b = r$ when the ratio including standard deviations for Y and X is 1. Therefore, when the analyses include standardized variables (e.g., $s_X = s_Y = 1$), a difference between correlations can be a useful measure of moderating effect because correlations equal regression coefficients. However, if this ratio is not 1, then b and r are not necessarily equal and, consequently, a difference between correlations may not provide a good indicator of the moderating effect.

Proportion of Variance Explained by the Moderating Effect as Indexed by R^2

As noted in Chapter 2, one way to examine whether the moderator variable has a statistically significant effect is to compare the R^2 from Model 2 (i.e., including the first-order effects and the product term) and Model 1 (i.e., including the first-order effects only). The difference between these two coefficients of determination (i.e., ΔR^2) indicates the proportion of variance in Y explained by the interaction effect above and beyond the proportion of variance explained by the first-order effects. Because this information is readily available in most computer outputs, ΔR^2 seems to be the most frequently used way to gauge the magnitude of the moderating effect.

What is a sufficiently large ΔR^2 to be considered "practically significant"? As noted in Chapter 3, the context and specific research questions provide an answer to this question. However, a review of 22 published studies of the job design literature has found that the median ΔR^2 is .02 (i.e., 2% variance explained) (Champoux & Peters, 1987). In addition, based on a Monte Carlo simulation, Evans (1985) concluded that "a rough rule would be to take 1% variance explained as the criterion as to whether or not a significant interaction exists in the model" (p. 320). This conclusion is based on the fact that, in Evans's simulation, when the population scores did include a moderating effect, sample-based results showed consistently that ΔR^2 was 1% or greater, whereas when the population scores did not include a moderating effect, results showed that ΔR^2 was typically smaller than 1%. In short, empirical and simulation results indicate that a statistically significant ΔR^2 of approximately .01 to .02 is an effect size worth taking seriously.

Proportion of Variance Explained by the Moderating Effect as Indexed by f^2

Although ΔR^2 seems to be the most widely used indicator of effect size, f^2 is a more accurate indicator because ΔR^2 is based on the proportion of Y variance explained by the product term vis-à-vis unexplained variance in Y after prediction based on the first-order terms (i.e., Model 1, Chapter 3). In contrast, f^2 is based on the proportion of Y variance explained by the product term vis-à-vis unexplained variance in Y after prediction based on the first-order terms and the product term. That is,

$$\Delta R^2 = \frac{R_2^2 - R_1^2}{1 - R_1^2} \qquad (9.1)$$

whereas

$$f^2 = \frac{R_2^2 - R_1^2}{1 - R_2^2} \qquad (9.2)$$

where R_1^2 is the proportion of variance in Y accounted for by the effects of X and Z (i.e., Model 1), and R_2^2 is the proportion of variance in Y accounted for by the effects of X, Z, and the product term $X \cdot Z$ (i.e., Model 2) (Aiken & West, 1991, p. 157).

Cohen et al. (2003, Chapter 5) suggested that effect sizes around f^2 = .02, .15, and .35 be labeled small, medium, and large, respectively. However, they offered the caveat that even effect sizes labeled "small" can have substantial practical and theoretical importance. Moreover, an effect size considered small in one field may be considered quite large in another, based on the consequences of the expected outcomes.

Proportion of Variance Explained by the Moderating Effect as Indexed by Modified f^2

The effect size indicator f^2 is not appropriate if the homogeneity of error variance is violated (see Chapter 4). Accordingly, Aguinis et al. (2003) developed a modified f^2 that is appropriate for situations including categorical moderator variables when there is heterogeneity of error variance. The formula for this modified f^2 is as follows (Appendix D includes further detail on the derivation of this equation):

$$\text{Modified } f^2 = \frac{\sum_{i=1}^{g}(n_i - 1)\rho_i^2\sigma_i^2 - \dfrac{\left[\sum_{i=1}^{g}(n_i - 1)\rho_i\sigma_i s_{X_i}\right]^2}{\sum_{i=1}^{g}(n_i - 1)s_{X_i}^2}}{\sum_{i=1}^{g}(n_i - 2)\ \sigma_i^2(1-\rho_i^2)} \qquad (9.3)$$

where n_i is the sample size in each moderator-based subgroup, ρ_i is the population X–Y correlation in each subgroup, σ_i^2 is the population Y variance for each subgroup, and $s_{X_i}^2$ is the X variance in each subgroup. Note that in computing modified f^2, the parameters ρ_i and σ_i^2 can be estimated by using their sample counterparts r_i and s_i^2.

Although the computation of the modified f^2 can be performed manually, a program written in Java and executable using any Internet browser is available at *www.cudenver.edu/~haguinis/mmr* to perform this computation based on input provided by the user. The needed input includes the following information: sample size in each subgroup, correlation between X and Y in each subgroup, Y variance in each subgroup, and X variance in each subgroup. The same considerations discussed earlier pertaining to what is a meaningful size for f^2 apply to the modified f^2 index.

Aguinis et al. (2003) reviewed all articles that used MMR to assess moderating effects of categorical variables published from 1969 to 1998 in *Journal of Applied Psychology, Personnel Psychology,* and *Academy of Management Journal*. This review resulted in a total of 106 articles, including 636 MMR analyses. Part of this review included the computation of the modified f^2 index for each of the MMR analyses. The resulting overall mean observed effect size was .009, with a 95% CI ranging from .006 to .012. Thus, the observed effect sizes as measured by the modified f^2 index were distributed around the .01 criterion derived by Evans (1985) for the ΔR^2 index. Based on Cohen et al.'s (2003) rules of thumb regarding effect size, $f^2 = .02$ is a small effect. Thus, the Aguinis et al. review revealed that the mean effect size in research published in three top-tier social science journals is close to this level. The resulting small modified f^2 is consistent with several literature reviews showing that small effect sizes are the norm across a large number of statistical tests and fields (e.g., Cohen, 1988; Mazen, Graf, Kellogg, & Hemmasi, 1987; Mazen, Hemmasi, & Lewis, 1987).

Measures of Improved Prediction

Differences in Unstandardized Regression Coefficients across Moderator-Based Subgroups

A method used to understand the practical significance of the moderating effect vis-à-vis improved prediction consists of comparing the unstandardized regression coefficients predicting Y from X across the moderator-based subgroups. As noted in Chapter 3, a separate regression equation can be derived for each moderator-based subgroup. In the fictitious data set predicting salary increase based on performance scores for tenured and untenured faculty, Chapter 3 showed that the following equations were derived for each of the groups:

Tenured Faculty: *Predicted* Salary = 183.25 + 385.62 *Perf*   (9.4)

Untenured Faculty: *Predicted Salary* = 155.45 + 105.42 *Perf*   (9.5)

This illustration shows that a 1-unit increase in performance is predicted to yield $385.62 for tenured faculty, but only $105.42 for untenured faculty. Moreover, Equations 9.4 and 9.5 show that the groups also differ in terms of the intercepts. Therefore, solving Equations 9.4 and 9.5 indicates that a tenured faculty member receiving a performance score of 4 (recall the performance scale ranges from 1 to 5) would receive a salary increase of approximately $1,726, whereas an untenured faculty also receiving a 4 on performance would receive a salary increase of approximately $577. Is this practically meaningful? The answer depends on the context. For example, is this difference in the relationship between performance and salary increase across groups sufficiently large that some promising junior (untenured) scholars may develop intentions to leave the university because they feel treated unfairly? Is this difference sufficiently large that, assuming there are more untenured than tenured female faculty members, a lawsuit may be filed by the female faculty on the grounds of pay inequity? Obviously, if the answer to these questions is yes, then the difference is sufficiently large to be taken seriously.

In the context of research situations, although a difference between regression coefficients may seem "small," the effect can be also practically significant. Consider, for instance, research on smoking cessation interventions. A small difference between slopes for the relationship between number of sessions attended and number of cigarettes smoked for Whites versus Latinos can have enormous public health implications given the approximately half a million annual deaths caused by tobacco consumption in the United States (Aguinis, Pierce, & Quigley, 1993).

Standardized Effect of the Moderator on the Y on X Slope

A second method used to assess relative improved prediction across moderator-based subgroups consists of computing the standardized effect of the moderator variable on the slope of Y on X (Champoux & Peters, 1987). This measure expresses the standardized rate of change in the slope associated with the moderator variable. Therefore, this measure can be used to compare the strength of a moderating effect across studies, or even in situations when an entirely different moderator variable is used.

The measure is computed as follows (Champoux & Peters, 1987, p. 248, Equation 4):

$$\beta_{\text{moderator effect}} = b_3 \sigma_z \frac{\sigma_X}{\sigma_Y} \qquad (9.6)$$

where b_3 is the unstandardized regression coefficient for the product term from the MMR model, and σ_Z, σ_X, and σ_Y are the population standard deviations for the moderator Z, the predictor X, and the criterion Y, respectively, across the entire sample. Note that Equation 9.3 provides the standardized effect of the moderator on the slope of Y on X because $b_3 \sigma_Z$ gives the change in slopes and σ_X/σ_Y standardizes the slopes. Sample-based s_Z, s_X, and s_Y can be used as estimates for the population standard deviations σ_Z, σ_X, and σ_Y. Recall that Chapter 5 noted that sample size across the moderator-based subgroups is not the same as the variance of the categorical moderator Z. For example, the variance of a binary moderator Z is (Aguinis, Boik, & Pierce, 2001):

$$s_z^2 = \frac{\Sigma(Z_i - \overline{Z})}{N-1} = \frac{Np(1-p)}{N-1} \qquad (9.7)$$

where $N = n_1$ (i.e., sample size in subgroup 1) + n_2 (i.e., sample size in subgroup 2), and $p = n_1/N$.

Two issues need to be noted. First, $\beta_{\text{moderator effect}}$ should not be confused with the "beta" printed in the "standardized" solution section of the output. Second, when X and Z are orthogonal (i.e., uncorrelated), $\beta_{\text{moderator effect}}$ is equal to ΔR (Champoux & Peters, 1987).

Once again, let us use the fictitious data set regressing salary increase on performance as moderated by tenure status. Results described in Chapter 3 indicate that the regression coefficient associated with the interaction between tenure and performance is $b_3 = -280.20$. This coefficient means that there is a -\$280.20 difference between the slope of salary increase on performance between the untenured (coded as 1) and the tenured group (coded as 0). In addition, results described in Chapter 3 indicate that, across the entire sample, the standard deviation for performance is $s_X = .91$ and the standard deviation for salary increase is $s_Y = 505.94$. Solving Equation 9.7 yields

$$s_z^2 = \frac{400(.4)(1-.4)}{400-1} = 96/399 = .24$$

Thus, solving Equation 9.6 yields

$$\beta_{\text{moderator effect}} = -280.20(.49)\frac{.91}{505.94} = -.25 \qquad (9.8)$$

The interpretation of this result is that the slope regression salary on performance is .25 standard deviations steeper for the tenured as compared to the untenured group.

Cohen (1988) suggested that differences expressed in standard deviations units can be considered small if they are around .20, medium if they are around .50, and large if they are around .80. Thus, extrapolating from Cohen's (1988) guidelines, the conclusion is that the moderating effect of tenure is a little larger than what is considered to be a small effect. Once again, however, a caveat must be made that the importance of a specific effect size must be considered in light of the context. The guidelines regarding what is a small, medium, and large effect size may not generalize to all situations. In fact, these guidelines were developed originally based on very specific bodies of literature. Specifically, Cohen's (1962) initial definitions of small, medium, and large effect size for various statistics were based on a literature review including all articles published in just one volume (i.e., 1961) of one journal (i.e., *Journal of Abnormal and Social Psychology*). For example, Cohen (1962) noted that "the level of average population proportion at which the power of the test was computed was the average of the sample proportions found" and "the sample values were used to approximate the level of population correlation of the test" (p. 147). Because Cohen's now-conventional definitions of small, medium, and large effect sizes are in part based on observed values, they have been revised over time as a consequence of subsequent literature reviews of effect sizes in various domains. For example, for correlation coefficients, Cohen defined .20 as small, .40 as medium, and .60 as large in his 1962 *Journal of Abnormal and Social Psychology* review. However, he changed these definitions to .10, .30, and .50, respectively, in his 1988 power analysis book. In short, the definitions of particular effect sizes as small, medium, and large have been based on observed effects is specific bodies of literature and should not be assumed to generalize to every research context. Moreover, Chapter 5 described a literature review concluding that moderating effects as observed in three social science journals are substantially smaller than Cohen's (1988) suggested standards.

Differential Impact of the Moderator at Various Values of the Predictor X

A third measure that helps our understanding of the practical significance of the moderating effect consists of examining the relative impact of the moderator variable at various levels of the predictor X (Aiken & West, 1991, pp. 132–133; Rogosa, 1980; Witt & Nye, 1998). This measure allows researchers to understand whether the difference in Y on X slopes across groups is more accentuated at the low, medium, or high score ranges for X.

Let us return to our illustrative data described in Chapter 3 for which tenure status was found to moderate the relationship between performance ratings and salary increase such that this relationship was stronger for tenured as compared to untenured faculty. The question answered by this measure of moderating effect importance is whether the moderating-effect of tenure status is stronger for low, medium, or high values of performance ratings.

Of course, we could see the relative impact of the moderator at various levels of X by plotting the slope of Y on X for each of the two groups. A perusal of Figure 9.1 shows that larger differences in predicted salary increase across the groups occur at the high end of the performance scale. As performance scores decrease, so does the difference in predicted salary increase across the groups. But, besides this general graphic-based information, we may wish to have a more precise understanding of such differences.

Once a moderator variable has been found (i.e., the coefficient associated with the product term is statistically significant), we can gain more precise understanding of differences in predicted salary across groups for specific values of X. Using the illustrative data set, we may be interested in knowing whether there is a large difference in predicted salary between tenured and untenured faculty for a performance score of 2 (i.e., poor performers) and 4 (i.e., high performers). The concern may be that if there are very large differences for top performers, the best untenured faculty may feel treated unfairly and choose to leave the university.

Recall that in the MMR model including first-order and interactive effects based on uncentered *Performance* is the following:

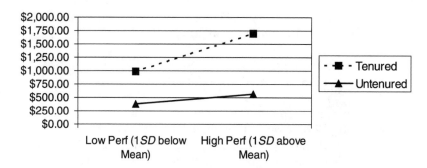

FIGURE 9.1. Slopes for *Salary* on *Performance* for tenured and untenured faculty based on the fictitious data set discussed in Chapter 3.

$$Predicted\ Salary = 183.25 + 385.62\ Perf - \tag{9.9}$$
$$27.80\ Tenure - 280.20\ Perf \cdot Tenure$$

As described in detail in Chapter 3, the interpretation of the regression coefficient for *Tenure* is that the estimated difference between the salary increase of an untenured faculty (coded as 0) and a tenured faculty (coded as 1), both with a performance score of 0, is −$27.80. However, a score of 0 does not even exist for *Performance* because it ranges from 1 to 5. Thus, the regression coefficient for *Tenure* in Equation 3.4 is not really meaningful. Therefore, Chapter 3 discussed the option of centering Performance scores (i.e., $CPerf = Perf - M_{Perf}$). After centering *Performance* and rerunning the analysis, the resulting equation is the following:

$$Predicted\ Salary = 1342.46 + 385.62\ CPerf - \tag{9.10}$$
$$870.10\ Tenure - 280.20\ CPerf \cdot Tenure$$

The interpretation of the regression coefficient associated with tenure is that the difference in salary increase between an untenured and a tenured faculty member with *average* performance scores (across the entire sample) is −$870.10.

A direct extension from the preceding centering procedure is to create a new, transformed performance score reflecting the point on the X scale in which we are interested. In other words, using the transformed performance variables $TPerf1 = Perf - 2$ and $TPerf2 = Perf - 4$ would yield regression equations with coefficients for *Tenure* indicating the difference between groups given that performance scores are 2 and 4, respectively. The resulting equations are the following:

$$\text{Predicted Salary} = 954.49 + 385.62 \ TPerf1 - \qquad (9.11)$$
$$588.19 \ Tenure - 280.20 \ TPerf1 \cdot Tenure$$

$$\text{Predicted Salary} = 1724.73 - 385.62 \ TPerf2 - \qquad (9.12)$$
$$1148.58 \ Tenure - 280.20 \ TPerf2 \cdot Tenure$$

The regression coefficient for Tenure in Equation 9.11 means that there is a \$588.19 difference in predicted salary increase in favor of the tenured group when the performance score is held constant at 2. The regression coefficient for Tenure in Equation 9.12 means that this difference is increased to \$1148.58 when performance score is 4. This result is not entirely surprising given Figure 9.1, but the implementation of this procedure allows for more precise comparisons in predicted salary increase across groups for specific points along the performance score continuum as compared to eyeballing the graph. These more precise results in turn provide additional information that is useful in gauging the practical significance of the moderator. In this particular example, we know that the difference in predicted salary increase is more than twice as large for high (i.e., score of 4 on a 1–5 scale) as opposed to low performers (i.e., score of 2). This is a practically significant result because the university would be particularly harmed if its best performing junior faculty leave because they feel they are treated unfairly vis-à-vis senior colleagues with similar performance levels.

Witt and Nye (1998) suggested a procedure to answer a related question: What is the point on the X scale where there is a standardized difference in predicted Y scores across the groups of a particular magnitude (i.e., small, medium, large)? Assume we wish to know the value on the X scale for which there is certain difference in predicted Y scores across groups. To answer this question, Witt and Nye (1998) provided the following equation:

$$X_{\text{expected difference}} = \frac{(d)(s_Y) - b_2}{b_3} \qquad (9.13)$$

where d is the expected difference between groups for predicted Y scores that one wishes to investigate, s_Y is the standard deviation of the criterion the entire sample, b_2 is the regression coefficient of the moderator variable (in the full model and based on the centered quantitative predictor), and b_3 is the regression coefficient associated with the product term. It is useful to express the resulting value for $X_{\text{expected difference}}$ in absolute score metric because the slopes across groups may cross at the high or low range of X. Thus, expressing the result in absolute score metric allows for easier interpretation.

Let us implement the procedure using the data set including information on the moderating effect of tenure status on the relationship between performance ratings and salary increase. First, let us first determine the X score at which a small difference of .20 (as defined by Cohen, 1988) is predicted for the salary increase across groups. Recall that the standard deviation for salary increase is $s_Y = 505.94$, and the MMR model including centered performance is shown in Equation 9.10. Solving Equation 9.13 for a small difference (i.e., $d = .20$) yields

$$\left| \frac{(.20)(505.94) + 870.10}{-280.20} \right| = 3.46$$

Now, let us compute the score at which a medium difference of .50 is found:

$$\left| \frac{(.50)(505.94) + 870.10}{-280.20} \right| = 4.01$$

The score at which a large difference of .80 exists is

$$\left| \frac{(.80)(505.94) + 870.10}{-280.20} \right| = 4.55$$

In other words, there is a small (as defined by Cohen, 1988) predicted difference in salary increase across groups for a performance score of 3.46, a medium-size difference for a performance score of 4.01, and a large difference for a performance score of 4.55 or higher. Of course, these results are consistent with Figure 9.1 as well as the procedure described earlier proposed by Aiken and West (1991) in that differences between groups are accentuated at the high end of the performance score scale. However, the Witt and Nye (1998) equation allows for the computation of the precise point on the X scale for which we expect to find a difference of a specific magnitude across groups.

Now, let us turn to two additional issues regarding the interpretation of moderating effects: the signed coefficient rule and the effects of swapping the criterion and predictor variables in the MMR model.

THE SIGNED COEFFICIENT RULE FOR INTERPRETING MODERATING EFFECTS

It is fairly common for MMR users to rely on an examination of the sign associated with the product term to make conclusions regarding which level of the moderator leads to greater predictability (e.g., Abdel-Halim,

1980; Bhagat, 1982; Fry, Kerr, & Lee, 1986). Specifically, considering the following typical MMR model including a binary moderator:

$$Y = a + b_1 X + b_2 Z + b_3 X \cdot Z + e \qquad (9.14)$$

Many authors have interpreted a positive sign for b_3 as meaning that high values of Z are associated with greater predictability, whereas a negative sign is interpreted as meaning that low values of Z are associated with greater predictability (Mossholder, Kemery, & Bedeian, 1990, provide several additional examples). However, in some situations using this "signed coefficient rule" (SCR) in isolation to interpret moderating effects can lead to misleading conclusions. Moreover, quite clearly, the sign for b_3 can only be interpreted given the coding scheme used for the moderator Z.

Consider the following simulation result involving a binary moderator variable (Mossholder et al., 1990). Holding standard deviations constant, when the $X–Y$ correlation for Subgroup 1 (coded as 1) is –.70 and the $X–Y$ correlation for Subgroup 2 (coded as 0) is .20, the resulting regression coefficient is –.62. The SCR indicates that there is greater predictability for the second subgroup because the resulting regression coefficient is negative and the second subgroup has the smaller value for Z (i.e., 0 as opposed to 1). However, the opposite conclusion is true because the correlation coefficient is greater (in absolute terms) for the first subgroup.

Recall than when dummy coding is used, b_3 represents the Y on X slope for the subgroup coded as 1 minus the Y on X slope for the subgroup coded as 0. Thus, in this example, the interpretation of a negative sign for b_3 based on the dummy coding system used indicates that the slope for Subgroup 1 is steeper than the slope for subgroup 2. This is an accurate interpretation of this result.

In short, the rather common practice of using the signed coefficient rule in isolation to interpret the nature and form of an interaction effect should be discontinued. This practice can lead to ambiguous and even erroneous interpretations regarding the nature of the interaction effect. Quite clearly, the interpretation of the sign for b_3 is only meaningful when the coding scheme chosen for the moderator Z is taken into account.

THE IMPORTANCE OF IDENTIFYING CRITERION AND PREDICTOR A PRIORI

In many social science areas interested in testing moderating effects, it is not clear whether Y is indeed the criterion or could be treated as a predictor, and whether X could also be treated as the predictor or the criterion. For example, in the labor economics literature, tests using moderators such as gender or ethnicity of the performance–wages relationship sometimes treat wages as the criterion and performance as a predictor (i.e., "direct" regression), whereas in other cases wages is treated as a predictor and performance is treated as the criterion (i.e., "reverse" regression) (White & Piette, 1998). The use of reverse regression has been proposed as a method leading to less bias in the estimation of the ethnicity- or gender-based discriminatory effect (Conway & Roberts, 1983), and has been used to study ethnicity-based discrimination in other social science areas such as mortgage-loan denial (LaCour-Little, 1996).

Another example of research domains in which the predictor and the criterion may be swapped is the organizational commitment literature, where job satisfaction is sometimes treated as a predictor of organizational commitment, whereas in others cases organizational commitment is treated as a predictor of job satisfaction (Landis & Dunlap, 2000). It is important that researchers make an a priori, and theory-based, distinction between the predictor and the moderator because results regarding the moderating effect of Z are likely to differ when the predictor and the criterion are swapped. In fact, Landis and Dunlap (2000) conducted a Monte Carlo simulation and found that in over 40% of the cases the conclusion about the presence of a moderating effect changed when the predictor and the criterion were swapped in the MMR model.

Let us use our data set to illustrate the importance of the a priori distinction between the predictor and the criterion. Our MMR model is the following:

$$Predicted\ Salary = a + b_1 Perf + b_2 Tenure + b_3 Perf \cdot Tenure \qquad (9.15)$$

Let us conduct the analysis using uncentered predictors because for this illustration we are not concerned about interpreting the regression coefficients associated with the first-order terms (i.e., b_1 and b_2). The resulting equation is the following:

$$Predicted\ Salary = 183.25 + 385.62\ Perf - \qquad (9.16)$$
$$27.80\ Tenure - 280.20\ Perf \cdot Tenure$$

As described in Chapter 3, the moderating effect of tenure resulted in an R^2 change of .062, $F(1, 396) = 292.79$, $p < .001$. The R^2 resulting from Equation 9.16 is .917.

Now, let us conduct an alternative analysis where the criterion *Sal-*

ary is treated as a predictor and the predictor *Performance* is treated as the criterion, yielding the following model:

$$\text{Predicted Perf} = a + b_1 Salary + b_2 Tenure + b_3 Salary \cdot Tenure \qquad (9.17)$$

Results for this new MMR model are the following:

$$\text{Predicted Perf} = .2580 + .0020\ Salary + .8980\ Tenure + \qquad (9.18)$$
$$.0019\ Salary \cdot Tenure$$

The moderating effect of tenure results in an R^2 change of .047, $F(1, 396) = 43.05$, $p < .001$. The R^2 corresponding to Equation 9.18 is .571. Thus, whereas the model shown in Equation 9.15 using *Salary* as the criterion results in an additional 6.2% of variance explained by the moderating effect of tenure, the model shown in Equation 9.17 results in additional variance of only 4.7% explained by the moderating effect. Also, the model including the product term explains a total of 91.7% of variance in *Salary*, whereas the model including the product term explains a total of 57.1% of variance in *Performance*.

In short, it is important to conceptualize the role of the criterion and the predictor in the MMR model. The criterion and predictor variables are not interchangeable, and swapping them can lead to discrepancies in the interpretation of the moderating effect. Researchers should be aware that a lack of clarity in the specification of the criterion and the predictor can lead to ambiguous conclusions regarding the role of the moderator variable.

CONCLUSIONS

■ This chapter discussed three topics regarding the interpretation of MMR results. First, a question often asked by MMR researchers and journal reviewers alike is whether a statistically significant moderating effect is also "practically significant." This chapter described several indicators that can be used to answer this question. These indicators allow MMR users to interpret results in terms of fit (i.e., proportion of variance in Y explained) or prediction (i.e., how the effect of X on Y changes across groups).

■ Regarding indicators of fit, the chapter discussed the use of differences in correlation coefficients across groups and the proportion of variance explained by the moderator as indexed by R^2, f^2, and modified f^2. Difference between correlations should be interpreted with caution because correlation and regression coefficients are the same for each group only when the ratio of Y to X is 1. When the ratios are not 1, differential validity (i.e., difference between correlations) does not yield the same result as differential prediction (i.e., difference between slopes). The use of ΔR^2 as an indicator of proportion or variance explained is quite pervasive possibly because of its wide availability in computer outputs and its transportability from study to study. In other words, proportion of variance explained does not depend on the metric used to measure any of the variables involved in the MMR model. In spite of the widespread use of ΔR^2, f^2 is a preferred index of proportion of variance explained because it refers to unexplained variance in Y after prediction based on the predictors and the product term; in contrast, ΔR^2 refers to proportion of variance explained in reference to unexplained variance in Y based on the first-order effects only. Moreover, the modified f^2 index is an optimal estimator of effect size because it can be used in situations where the homogeneity of error variance assumption has been violated.

■ Regarding indicators of improved prediction, the chapter discussed the use of differences in unstandardized regression coefficients across groups, the standardized effect of the moderator on the slope of Y on X, and the differential impact of the moderator at various values of the predictor X. An examination of differences between unstandardized regression coefficients provides researchers with a good understanding of differential predictability across groups for any given study. However, unstandardized coefficients are referenced to the specific scales used in each study and, therefore, it may be difficult to compare such differences across studies when different measures are involved. This is why meta-analysts usually prefer to accumulate correlation coefficients as opposed to regression coefficients across studies (Kanetkar, Evans, Everell, Irvine, & Millman, 1995; Raju, Fralicx, & Steinhaus, 1986; Raju, Pappas, & Williams, 1989). The standardized effect of the moderator on the slope of Y on X overcomes this potential problem because it allows researchers to compute the difference in slopes across groups in standard deviation units. Finally, the chapter described how researchers can calculate the impact of the moderator at various levels of the predictor X as well as the point on the X scale where there is a standardized difference in predicted Y scores of a specific magnitude. Although these are not effect size indexes in the traditional sense, they provide additional information to help researchers understand the importance of the moderating effect.

■ Ultimately, the answer to the question regarding practical significance depends on the research context and the practical and theoretical importance of the question that is investigated. Nevertheless,
the indicators discussed in this chapter provide MMR researchers
with a tool kit to better understand whether their MMR results are
likely to have practical impact. The gathering of evidence of whether
a moderating effect is "practically significant" is much like the work
of a detective. Thus, it is desirable to use more than one indicator
because the larger the accumulated evidence, the stronger the case
regarding the practical importance of the moderating effect.

■ A second topic discussed in this chapter regarding the interpretation
of moderating effects is the often-used "signed coefficient rule." This
rule can be misleading when used in isolation. MMR results should
be interpreted in light of the coding scheme that was chosen. Thus,
the practice of using the signed coefficient rule as a sole means to interpret moderating effects should be discontinued.

■ Finally, the chapter also addressed the fact that in some social science areas in which MMR is used there is not a clear rationale for
identifying which variable is the criterion and which one the moderator in an MMR analysis. However, this chapter showed that it is important to conceptualize the role of the criterion and the predictor.
The criterion and predictor variables are not interchangeable, and
swapping them can lead to discrepancies in the interpretation of the
moderating effect. In some cases, a moderating effect may be completely eliminated by swapping criterion and predictor. In others, a
moderating effect may emerge as a consequence of this swapping.
Therefore, MMR users should strive to always identify the criterion
and predictor variables a priori because failure to do so can lead to
discrepant results regarding the presence of the moderating effect.

■ The next and final chapter provides a summary of the previous
chapters in the volume and offers some guidelines and recommendations for the use and interpretation of MMR to estimate moderating effects of categorical variables.

10

Summary and Conclusions

Don't let it end like this. Tell them I said something.
—PANCHO VILLA

This chapter integrates and provides a summary of the main points
discussed in each of the chapters included in the volume. In addition, it offers specific guidelines and recommendations for the use
and interpretation of MMR to estimate moderating effects of
categorical variables.

MODERATORS AND SOCIAL SCIENCE THEORY AND PRACTICE

Moderated or interactive relationships are at the heart of the scientific
enterprise because they provide information on when or under which
conditions the relationship between two variables is likely to change.
This statement echoes the sentiment of numerous social scientists that
researchers must have a good understanding of the moderator variables
in their fields. A lack of understanding regarding moderators is likely to
lead to a delay in theory development and interventions that lead to
unintended consequences.

From a theory standpoint, a thorough understanding of moderators allows researchers to learn whether causal relationships generalize
or whether there are boundary conditions for the model in question. Because of their importance for theory advancement, moderator variables
play a central role in social science research. Some examples include the
debates of whether personality consistency is moderated by individual
differences, whether the relationship between preemployment test
scores and performance criteria is moderated by gender and ethnicity,

155

and whether the relationship between instructional techniques and learning is moderated by student characteristics, to name just a few.

Because moderators play such a central role in social science theory, it follows that they are also critical in social science practice. A good understanding of moderators allows researchers to plan and execute interventions with the greatest impact and effectiveness. For instance, does the implementation of a smoking cessation program lead to the same degree of success rates for members of various ethnic groups? Not having an accurate answer to this question may lead to the implementation of a smoking cessation program that is highly effective for one group but virtually ineffective for another.

Researchers should be aware of the importance of having a strong justification before launching into a moderator analysis. Going on a "fishing expedition" in search for moderators is likely to lead to conclusions that are not replicable in subsequent studies. In some cases, researchers may fail to find support for existing moderators. For instance, Villa, Howell, Dorfman, and Daniel (2003) argued that the disregard for theory is one of the chief reasons why researchers have not found support for moderators in leadership research. Specifically, they noted that testing relationships (whether or not they are hypothesized) and including variables in the regression equation (whether or not there is a logical reason to be included in the model) are among the chief reasons why leadership researchers have been unable to find support for moderated relationships. In yet other cases, the lack of regard for theory may lead researchers to find "false" moderators. However, in the majority of cases, researchers are likely to incorrectly conclude that there is no moderating effect.

USE OF MODERATED MULTIPLE REGRESSION

Moderated multiple regression (MMR) is an extension of OLS regression. The difference is that the MMR model incorporates an additional predictor carrying information regarding the moderating effect. The test of the moderating effect consists of assessing whether the regression coefficient associated with the product term between the predictor and the moderator is different from zero in the population. This test is conducted by computing a t statistic. Alternatively, and equivalently, researchers can assess whether the inclusion of the product term in the regression equation including the first-order effects of the predictors improves the ability to account for variance in the criterion. This test is conducted by computing the difference between the R^2 for Model 2 (i.e., first-order effects and product term) and Model 1 (i.e., first-order effects only). Then an F statistic is computed for the resulting ΔR^2. The significance level for the t statistic associated with the regression coefficient for the product term and the F statistic associated with ΔR^2 are identical. In other words, the conclusion of whether a moderating affect exists is identical regardless of which statistical test is used. An examination of the regression coefficient associated with the product term provides information on how much the slope of Y on X differs across moderator-based subgroups. On the other hand, ΔR^2 provides information on whether the moderating effect explains additional variance in the criterion above and beyond the effects of the first-order effects.

MMR has been endorsed by numerous independent analytic and empirical investigations as well as reports and guidelines issued by social science professional organizations. Likely as a consequence of these endorsements, MMR is used pervasively in social science research. It is probable that every year there are hundreds of MMR analyses including categorical variables published in social science journals. In fact, MMR seems to be the method of choice for estimating moderated relationships including categorical moderator variables in a variety of fields.

MMR can be conducted using any of the widely available statistical software packages. Chapter 3 provided a step-by-step description of how to conduct an MMR analysis including a binary moderator variable. Chapter 8 provided an extension to cases including moderator variables with more than two levels. The procedure involves creating a new variable that consists of the product term between the predictor and the moderator variables, and then implementing a hierarchical regression procedure. The computer output includes information that allows researchers to answer the key issue of whether the slope of the criterion on the predictor differs across the moderator-based subgroups.

It is recommended that researchers use dummy coding when the moderator includes two levels only. In this type of coding, members of one of the groups are assigned a 0 and members of the other group are assigned a 1. This coding scheme is recommended for binary moderators because of its simplicity and ease of interpretation of the results. The choice of a coding scheme affects the interpretation of the intercept and the regression coefficients in the MMR model. Coding schemes other than dummy coding represent a different way of partitioning the total variance associated with the interaction. Therefore, the intercept and the regression coefficients for the first-order effects answer a different question for each coding scheme and, consequently, the interpretation of results needs to take the coding scheme into account. On the other hand, the significance level for the coefficient for the product term as well as the R^2s for Model 1 and Model 2 are not affected by a change in the coding scheme.

Centering of the quantitative predictor X is an additional issue that should be taken into account if a researcher wishes to interpret the intercept and the first-order effect coefficients in the presence of a nonzero interaction. When the quantitative predictor X is not centered, the intercept and the coefficient for the first-order effect of the predictor hypothesized to be the moderator are referenced to a *zero* point on the X, and this zero point may not be meaningful given the scale used to measure X. On the other hand, the intercept and coefficient resulting from centering the X variable is referenced to the *average* value on X. Thus, centering X is likely to result in a more meaningful intercept and regression coefficient for the first-order effect of the moderator. Note, however, that the size, statistical significance, and interpretation of the coefficient associated with the product term are identical in equations based on centered and uncentered predictors.

In situations including moderator variables with more than two levels (i.e., $k > 2$), researchers need to first create $k - 1$ code variables. Then product terms are formed between the quantitative predictor and each of the code variables. In the case of a moderator with three levels, there are two code variables (i.e., Z_1 and Z_2). In this case, the MMR model includes the predictors X, Z_1, and Z_2 as well as $X \cdot Z_1$ and $X \cdot Z_2$. In such a model, Z_1 and Z_2 (taken together) represent the first-order effect of the predictor Z hypothesized to be the moderator and $X \cdot Z_1$ and $X \cdot Z_2$ form a set that represents the moderating effect of Z (i.e., interaction between X and Z). As in the case of binary moderators, choosing the coding scheme is a key consideration because the choice leads to the attachment of different meanings to each of the code variables. Chapter 8 included a discussion of the following four types of coding schemes: dummy coding, unweighted effect coding, weighted effect coding, and contrast coding. Theory and the expected form of the interaction effect are the key factors that dictate which coding scheme should be preferred.

Finally, the MMR model generalizes easily to situations including three or more predictors that may be hypothesized to interact with each other. All the considerations regarding interpretation of lower-order terms as well as centering, as discussed in Chapters 3 and 8, are relevant. That is, lower-order terms need to be entered in the equation, and the interpretation of lower-order coefficients is affected by the presence of a higher-order interaction effect. Thus, unless quantitative predictors are centered, the interpretation of lower-order effects may be virtually meaningless. Chapter 8 also addressed situations where researchers may wish to test for the presence of a targeted interaction effect (e.g., two-way interaction) in the presence of higher-order interaction effects (e.g., three-way interaction). The recommendation is to follow the logic that a researcher first examines the omnibus effect (i.e., three-way inter-action) and then proceeds to examine specific, more in-depth, compari-sons.

HOMOGENEITY OF ERROR VARIANCE ASSUMPTION

The homogeneity of error variance across levels of the moderator is a critical assumption of the MMR model including categorical modera-tors. Results of a selective literature review described in Chapter 4 show that the assumption is violated in approximately 50% of published stud-ies. This is an unfortunate finding because violating the assumption may lead to erroneous substantive conclusions. In turn, these erroneous substantive conclusions are likely to lead to erroneous theory and ineffectual social science interventions.

Satisfying the homogeneity of error variance assumption means that the residual terms from predicting Y from X are similar across the moderator-based populations. The homogeneity of error variance as-sumption is not the same as the more familiar homoscedasticity assump-tion (i.e., constant distribution of scores throughout the regression line). The homogeneity of error variance assumption can be violated even if the usual homoscedasticity is met for all the scores combined or for each of the moderator-based subgroups. The assumption is important because its violation can make MMR's F test results erratic. Type I error rates can be inflated, which leads researchers to discover a "false" moderator. Alterna-tively, Type I error rates can be overly conservative, which implies a decrease in statistical power to unacceptably low levels and incorrectly concluding that there is no moderation. Therefore, when the assumption is violated, substantive research conclusions are likely to be erroneous. In short, MMR's F test can be misleading and, therefore, should not be used in the absence of homogeneity of error variance.

Chapter 4 described the computer program ALTMMR, which is available at *www.cudenver.edu/~haguinis/mmr*, that allows for the assess-ment of whether the assumption has been violated. The program also allows for the computation of alternative statistics such as A and J that can be used in lieu of MMR's F test to ascertain whether there is a mod-erating effect when the assumption is violated.

LOW STATISTICAL POWER AND PROPOSED REMEDIES

Chapter 5 provided a detailed discussion of the many factors that have a detrimental effect on the statistical power of MMR. These factors are related to variable distributions (e.g., predictor variable variance reduction), operationalizations of criterion and predictor variable (i.e., measurement error, scale coarseness, polichotomization of truly continuous variables), sample size (e.g., sample size across moderator-based subgroups), and characteristics of the predictor variables (e.g., correlation between the predictor X and the criterion Y). In one way or another, each of these factors lowers the observed effect size as compared to the effect in the population. An underestimation of effect size causes statistical power to drop. Because a subset of these factors is present in most social science research situations, it is important to be aware of what they are and the mechanisms through which they affect power. Moreover, the factors that affect power downwardly also have interactive effects. This means that it is sufficient to have an adverse value on just one or two of the factors for power to decrease substantially below the suggested value of .80. A 30-year review of three high-quality social science journals described in Chapter 5 showed that the median observed effect size is .002. With such small observed effect sizes, and subsequently lower power, it is no surprise that moderators are hard to find!

The low power of MMR has important implications for both theory and practice. Regarding theory, it is likely that low power has led to the incorrect conclusion that there are no moderating effects in numerous research domains. This implication can open whole new research avenues. It may be useful to revisit published articles that reported no moderating effect in spite of well-grounded theoretical rationale and replicate these studies with MMR tests with greater power. We should not be surprised if predicted moderator variables are found given a more appropriate power level. From a practical standpoint, low power means that an intervention that does not have the same effect across groups is nevertheless used uniformly. Such erroneous implementation of social science interventions (e.g., preemployment testing, clinical treatment, smoking cessation programs, and so forth) is likely to lead to uneven desired outcomes across groups (e.g., predicted job performance, mental health, smoking cessation success rate). Low power can lead to the unexpected result that a particular intervention did not work as well for all groups under consideration. Such a finding has implications in terms of fairness for the participants as well as the optimization of resources used in implementing the intervention.

Although there are numerous factors that affect the power of MMR adversely, MMR users can make research design and measurement choices that enhance statistical power. Chapter 6 described several strategies available to address each of the factors that affect power. Some of these strategies are rather obvious, such as the use of computer programs described in Chapter 7 to plan research design so that sample size is sufficiently large, but not larger than needed, to detect the expected effect size. However, other strategies are less conventional, such as the use of a larger than traditional preset Type I error rate based on a consideration of the relative seriousness of committing a Type I versus a Type II statistical error. Also, not all strategies are available in every research situation. However, implementing as many of the available strategies as possible is an important consideration that should be taken seriously by all MMR users who wish to minimize the impact of the factors that are likely to lead to the incorrect conclusion that there is no moderating effect. Furthermore, researchers should not wait until the data have been collected to think about statistical power. On the contrary, the implementation of strategies that enhance power should begin at the research planning and design stages of conducting any study. Finally, researchers should be aware that the following is the rank order of factors affecting power in order of importance: (1) moderating effect magnitude, (2) total sample size, (3) sampling restriction on X, and (4) measurement error. This information is useful because it points to the factors that researchers should give priority in their attempts to maximize statistical power.

In addition to being aware of the factors that affect power and implementing strategies to minimize the impact of such factors, researchers are advised to compute power before a study is conducted. Conducting a power analysis helps researchers make decisions regarding various design alternatives vis-à-vis the resources available. For example, researchers can assess whether a specific increase in sample size leads to an increase in power sufficiently substantive to justify the time and cost associated with the additional data collection effort. Also, given the increased criticism of null hypothesis significance testing, numerous journal editors as well as funding agencies now require a power analysis routinely. Thus, it will become increasingly difficult to publish empirical research, particularly in the case of null findings, and request research funds in the absence of a power analysis.

Three computer programs are available at *www.cudenver.edu/ ~haguinis/mmr* to calculate the statistical power of MMR given specific research conditions. These programs were described in Chapter 7. Two of these programs, POWER and MMRPWR, are based on previously published empirical results, and the third one, MMRPOWER, is based on a theory-based approximation to power. These computer programs allow researchers to compute power in planning their research design, to assess the relative impact on power resulting from varying the factors that affect power, to conduct a cost–benefit analysis of the relative advantages and disadvantages of manipulating the factors that affect power in an actual study, and to assess the power of MMR analyses reported in published studies. Given the ease of use and availability of these programs, it is no longer justified to conduct an MMR analysis without an a priori statistical power assessment. It is also no longer justified to conclude that the null hypothesis of no moderating effect is correct unless a power assessment has resulted in a value sufficiently high to detect a hypothesized moderating effect.

MORE COMPLEX MMR MODELS

In the past few years several articles have been published regarding the difficulty in distinguishing moderating from nonlinear effects. This body of literature has led to the question of whether quadratic (i.e., nonlinear) terms should be included routinely in MMR models. The discussion provided in Chapter 8 led to the conclusion that quadratic effects should be included in MMR models when the following conditions are met: There is a clear theory-based rationale to include quadratic terms in the regression equation; correlation between first-order predictors is high; reliability of first-order predictors is low; and sample size is large. On the other hand, routinely including quadratic terms in the model in the absence of specific theory-based hypotheses is consistent with exploratory as opposed to confirmatory research. Results of the former should be taken with great caution. Such results must be replicated using independent samples, particularly if the model includes unexpected and complex nonlinear effects. As noted in other sections of this volume, there is no statistical technique that can substitute for good thinking. Thus, a key factor in deciding whether to include quadratic terms in an MMR model is whether there is a specific theory-based rationale for including such terms.

ASSESSING PRACTICAL SIGNIFICANCE

A question often asked by researchers is whether a statistically significant moderating effect is also "practically significant." Chapter 9 described several indicators that can be used to answer this question. These indicators allow MMR users to interpret results in terms of fit (i.e., proportion of variance in Y explained) or prediction (i.e., how the effect of X on Y changes across groups).

Regarding indicators of fit, Chapter 9 included a description of the use of differences in correlation coefficients across groups and the proportion or variance explained by the moderator as indexed by R^2, f^2, and modified f^2. The difference between correlations across groups should be interpreted with caution because correlation and regression coefficients are the same for each group only when the ratio of Y to X variances is 1. Therefore, when the ratios are not 1, the difference between correlations does not yield the same result as the difference between slopes. The use of ΔR^2 as an indicator of proportion or variance explained is seen quite frequently in published research possibly because it is reported in most computer software outputs. Also, proportion of variance explained does not depend on the metric used to measure any of the variables involved in the MMR model (although it does depend on the number of predictors included in the model). In spite of the widespread use of ΔR^2, f^2 is a preferred index of proportion of variance explained because it refers to unexplained variance in Y after prediction based on the predictors and the product term; in contrast, ΔR^2 refers to proportion of variance explained in reference to unexplained variance in Y based on the first-order effects only. Moreover, the *modified* f^2 index is an even more optimal estimator of effect size because it can be used in situations where the homogeneity of error variance assumption has been violated. The modified f^2 index can be computed by hand or by using a computer program available at *www.cudenver.edu/~haguinis/mmr.*

Regarding indicators of improved prediction, Chapter 9 discussed the use of differences in unstandardized regression coefficients across groups, the standardized effect of the moderator on the slope of Y on X, and the differential impact of the moderator at various values of the predictor X. An examination of differences between unstandardized regression coefficients provides researchers with a good understanding of differential predictability across groups for any given study. However, unstandardized coefficients are referenced to the specific scales used in each study and, therefore, it may be difficult to compare such differences across studies when different measures are involved. The standardized effect of the moderator on the slope of Y on X overcomes this potential problem because it allows researchers to compute the difference in slopes across groups in standard deviation units. Finally, Chapter 9 described how researchers can calculate the impact of the moderator at various levels of the predictor X as well as the point on the X scale where there is a standardized difference in predicted Y scores of a specific magnitude. Although these are not effect size indexes in the traditional sense, they provide additional information to help researchers understand the importance of the moderating effect.

Ultimately, the answer to the question regarding practical significance depends on the research context and the practical and theoretical importance of the question that is investigated. Nevertheless, the indicators discussed in Chapter 9 provide MMR researchers with a tool kit to better understand whether their MMR results are likely to have meaningful impact.

CONCLUSIONS

■ In spite of the widespread interest in moderator effects in the social sciences, researchers have repeatedly expressed the concern that moderators are elusive. This book provides a systematic description of how to use and interpret results about categorical moderators from multiple regression. Ideally, the information presented in this volume will improve the chances that researchers will detect hypothesized moderating effects.

■ There are some key issues that researchers must take into consideration in their quest for moderating effects. First, theory and good logic need to take precedence over research design and data analysis considerations. It is unlikely that, even if all the recommendations provided in this volume regarding the correct use and interpretation of MMR are implemented, researchers will be able to make meaningful conclusions regarding moderating effects if the model is not based on sound logic and theory. No research design or data analysis considerations will be able to compensate for the lack of good logic and theory.

■ Second, the statistical power of MMR needs to be computed *before* a study is conducted. Sometimes the best decision may be *not* to conduct a study because power will be too low. Results of literature reviews described in this book demonstrate that, in the vast majority of cases, research published in some of the top social science journals has lacked sufficient statistical power to detect moderating effects. Thus, it is imperative that researchers conduct an a priori power analysis to know whether their anticipated research design characteristics (e.g., sample size, error of measurement in the scales to be used) will be adequate for MMR to yield accurate results. If the power analysis shows insufficient levels of power, then corrective action needs to be taken before the data are collected.

■ Third, once the data are collected, researchers must be aware that the coding scheme used for the categorical moderator variable has profound implications on the interpretation of the moderating effect. Although statistical significance values for the moderating effect are not affected by the various coding schemes, the coding scheme affects the interpretation of the effect.

- Fourth, implementing additive transformations on the quantitative predictor also have implications regarding the interpretation of results. Specifically, the interpretation of the intercept and the regression coefficient associated with the moderator is in many cases not meaningful unless the quantitative predictor is transformed using centering or other transformations. However, there is no need to implement such transformations on the criterion or the categorical moderator.

- Fifth, researchers should check that their data sets comply with the homogeneity of error variance assumption. If this assumption is violated, MMR results can be misleading. Thus, when violation occurs, researchers are advised to implement alternative procedures to test for the presence of the moderating effect as described in Chapter 4.

- Sixth, if a moderating effect is found, there are several procedures available to better understand whether the moderating effect is practically significant. Ultimately, the issue of practical significance depends on the context and outcomes of a particular research study. However, indicators such as the modified f^2 effect size index provide information regarding the strength of the effect.

- In closing, MMR is like a minefield in the sense that there are multiple threats, many unknown to researchers, that are likely to affect the accuracy of conclusions regarding the presence of the moderating effect. These threats are present at each of the stages of a study from conceptualization (e.g., lack of clarity regarding moderation vs. mediation hypothesis) to the research design (e.g., insufficient statistical power), data analysis (e.g., use of centered vs. raw scores), and interpretation (e.g., conclusions regarding the first-order effect of the moderator based on using a specific coding scheme). Ideally, this volume will allow researchers to be aware of, anticipate, and minimize the impact of most such threats. A better knowledge of moderators will no doubt help advance social science theory and practice.

Appendix A

Computation of Bartlett's (1937) M Statistic

Bartlett's (1937) M statistic is approximately distributed as chi-square with $k - 1$ degrees of freedom when sample size in each of the moderator-based subgroups $n_k - 1 \geq 3$. Given that k = number of subgroups, n_k = number of observations in each subgroup, s^2 = subgroup variance on the criterion, and v = degrees of freedom from which s^2 is based, the M statistic is computed as follows:

$$M = \frac{(\sum_i v_i) \log_e (\sum_i v_i s_i^2 / \sum_i v_i) - \sum_i v_i \log_e s_i^2}{1 + \frac{1}{3(k-1)}(\sum_i 1 / v_i - 1 / \sum_i v_i)} \quad \text{(A.1)}$$

For unconditional subgroup variances, substituting $\sigma_{e_i}^2$ from Equation 4.1 for s_i^2 yields:

$$M = \frac{(\sum_i v_i) \log_e (\sum_i v_i \sigma_{e_i}^2 / \sum_i v_i) - \sum_i v_i \log_e \sigma_{e_i}^2}{1 + \frac{1}{3(k-1)}(\sum_i 1 / v_i - 1 / \sum_i v_i)} \quad \text{(A.2)}$$

Appendix B

Computation of James's (1951) *J* Statistic

To test for differential slopes, *J* is computed using Equation B.1 (Alexander & Govern, 1994):

$$J = \sum_{i=1}^{k} \left[\frac{(b_i - b^+)^2}{s_{b_i}^2} \right] \tag{B.1}$$

The various components of Equation B.1 are computed by implementing the following steps:

1. Determine the squared standard error ($s_{b_i}^2$), where

$$s_{b_i}^2 = \frac{(1 - r_i^2)s_{Y_i}^2}{(n_i - 2)s_{X_i}^2}.$$

2. Define a weight for each regression weight (i.e., b_i) such that $\sum w_i = 1$:

$$w_i = \frac{1 / s_{b_i}^2}{\displaystyle\sum_{i=1}^{k} 1 / s_{b_i}^2}.$$

3. The variance-weighted estimate of the common regression slope (b^+) then becomes $b^+ = \displaystyle\sum_{i=1}^{k} w_i b_i$.

Once the *J* statistic has been computed, the adjusted critical value (*c*) for a chi-square distribution with $k - 1$ degrees of freedom and nominal Type I error α is determined as follows:

1. Let $v_i = n_i - 2$.

2. $R_{st} = \sum_{i=1}^{k} \left[\dfrac{w_i^t}{v_i^s} \right]$ (note that values for s and t are only those re-

quired in the calculation of $h(\alpha)$ below).

3. $\chi_{2s} = \dfrac{c^s}{\prod\limits_{q=1}^{s}(k+2s-3)}$ where $\prod\limits_{q=1}^{s}$ denotes the product of each

term from 1 to s (note that s represents the multiplier to provide the values of χ_{2s}, [i.e., for χ_2, χ_4, χ_6, χ_8, s is 1, 2, 3, 4, respectively]).

4. $T = \sum_{i=1}^{k} \dfrac{(1-w_i)^2}{v_i}$

5. $h(\alpha)$ is calculated as follows:

$h(\alpha) = c + \tfrac{1}{2}(3\chi_4 + \chi_2)T +$

$$\left\{ \begin{array}{l} \left[(\tfrac{1}{16})(3\chi_4+\chi_2)^2 \dfrac{1-(k-3)}{c}T^2 + (\tfrac{1}{2})(3\chi_4+\chi_2) \times \right. \\[4pt] \left. \begin{bmatrix} (8R_{23}-10R_{22}+4R_{21}-6R_{12}^2+8R_{12}R_{11}-4R_{11}^2)+ \\ (2R_{23}4R_{22}+2R_{21}-2R_{12}^2+4R_{12}R_{11}-2R_{11}^2)\times(\chi_2-1)+ \\ (\tfrac{1}{4})(-R_{12}+4R_{12}R_{11}-2R_{12}R_{10}-4R_{11}^2+4R_{11}R_{10}-R_{10}^2)\times \\ (3\chi_4-2\chi_2-1) \end{bmatrix} \right] + \\[8pt] (R_{23}-3R_{22}+3R_{21}-R_{20})(5\chi_6+2\chi_4+\chi_2)+ \\ (\tfrac{3}{16})(R_{12}^2 4R_{23}+6R_{22}-4R_{21}+R_{20})\times \\ (35\chi_8+15\chi_6+9\chi_4+5\chi_2)+ \\ (\tfrac{1}{16})(-2R_{22}+4R_{21}-R_{20}+2R_{12}R_{10}-4R_{11}R_{10}+R_{10}^2)\times \\ (9\chi_8-3\chi_6-5\chi_4-\chi_2)+ \\ (\tfrac{1}{4})(R_{22}+R_{11}^2)(27\chi_8+3\chi_6+\chi_4+\chi_2)+ \\ (\tfrac{1}{4})(R_{23}-R_{12}R_{11})(45\chi_8+9\chi_6+7\chi_4+3\chi_2) \end{array} \right\}$$

The null hypothesis (i.e., $H_0: \beta_1 = \ldots = \beta_k$) is rejected when $J > h(\alpha)$.

Appendix C

Computation of Alexander's (Alexander & Govern, 1994) A Statistic

Calculation of the A statistic is similar to the computation for J described in Appendix B. To test for differences in slopes, the A statistic is calculated using Equation C.1 and referenced to the chi-square distribution with $k - 1$ degrees of freedom:

$$A = \sum_{i=1}^{k} z_i^2 \qquad (C.1)$$

where the following steps are required:

1. Determine the squared standard error ($s_{b_i}^2$), define a weight for each regression weight (i.e., b_i), and determine the variance-weighted estimate of the common regression slope (b^+) as in Steps 1–3 in Appendix B.

2. Define a one-sample t statistic for each subgroup where
$$t_i = \frac{b_i - b^+}{s_{b_i}}.$$

3. Square each t statistic and transform it by calculating z_i, where
$$z_i = c + \frac{(c^3 + 3c)}{b} - \frac{(4c^7 + 33c^5 + 240c^3 + 855c)}{(10b^2 + 8bc^4 + 1000b)} \quad \text{and}$$

$$a = v_i - .5 \qquad b = 48a^2 \qquad c = \sqrt{[a\ln(1 + t_i^2 / v_i)]}$$

where $v_i = n_k - 2$.

4. Square these z_is and sum to determine A as shown in Equation C1, and reference it to the chi-square distribution with $k - 1$ degrees of freedom.

The null hypothesis (i.e., $H_0: \beta_1 = \ldots = \beta_k$) is rejected when $A > h(\alpha)$.

Appendix D

Computation of Modified f^2

The hypothesis of no moderating effect in MMR is tested by comparing to $F_m = [\text{SSI}/k - 1]/[\text{SSE}/(N - 2k)]$ to $F_{k-1,N-2k}^{1-\alpha}$, where k is the number of moderator-based subpopulations, N is the total sample size (across all groups), SSI is the sum of squares due to the interaction between the categorical moderator variable Z and the continuous predictor X, and SSE is the error sum of squares after fitting first-order effects and the product term.

It can be shown that conditional on X

$$E(SSI) = \sum_{i=1}^{k} \sigma_i^2 (1 - \rho_i^2)(1 - w_i) + \sum_{i=1}^{k} (n_i - 1)\rho_i^2 \sigma_i^2 - \qquad \text{(D.1)}$$

$$\frac{\left[\sum_{i=1}^{k} (n_i - 1)\rho_i^2 \sigma_i^2 s_{x_i}\right]^2}{\sum_{i=1}^{k} (n_i - 1)s_{x_i}^2}$$

and

$$E(SSE) = \sum_{i=1}^{k} (n_i - 2)\sigma_i^2 (1 - \rho_i^2)$$

where $w_i = (n_i - 1)s_{x_i}^2 \Big/ \sum_{i=1}^{k} (n_j - 1)s_{x_j}^2$ is the variance of the observable Y scores in the ith subpopulation, $s_{x_i}^2$ is the sample variance of X in the

From Aguinis, Beaty, Boik, and Pierce (2003).

171

ith subpopulation, n_i is the sample size from the ith subpopulation, and σ_i is the correlation between Y and X in the ith subpopulation.

The first term in $E(SSI)$ reflects the $k - 1$ degrees of freedom associated with SSI. The remaining terms in $E(SSI)$ reflect variation accounted for by the categorical moderator by continuous predictor variable interaction. Accordingly, the effect size is as follows:

$$\text{Modified } f^2 = \frac{E(SSI) - \sum_{i=1}^{k} \sigma_i^2 (1 - \rho_i^2)(1 - w_i)}{E(SSE)} =$$

$$\frac{\sum_{i=1}^{k} (n_i - 1)\rho_i^2 \sigma_i^2 - \dfrac{\left[\sum_{i=1}^{k} (n_i - 1)\rho_i \sigma_i s_{x_i}\right]^2}{\sum_{i=1}^{k} (n_i - 1)s_{x_i}^2}}{\sum_{i=1}^{k} (n_i - 2)\sigma_i^2 (1 - \rho_i^2)}$$

If ρ_i is written as $\rho_i = \beta_i s_{x_i} / \sigma_i$, then the effect size can be written as

$$\text{Modified } f^2 = \frac{\sum_{i=1}^{k} w_i (\beta_i - \bar{\beta})^2}{\theta - \bar{\beta}^2 - \sum_{i=1}^{k} w_i (\beta_i - \bar{\beta})^2} + O\left(\frac{1}{N}\right), \qquad \text{where}$$

$$\theta = \frac{\sum_{i=1}^{k} (n_i - 1)\sigma^2}{\sum_{i=1}^{k} (n_i - 1)s_{x_i}^2}, \quad \bar{\beta} = \sum_{i=1}^{k} w_i \beta_i,$$

and w_i is defined in Equation D.1. Note that if $\beta_i = \rho_i \sigma_i / s_{x_i}$ is constant for all i (i.e., $\beta_i = \bar{\beta}$), the $f^2 = 0$. To estimate effect size, sample quantities can be substituted for population parameters.

Appendix E

Theory-Based Power Approximation

The power of the MMR F test is

$$Power \approx Pr\left[\left(\frac{k-1}{N-2k}\right)F_{k-1,N-2k}^{1-\alpha}\sum_{j=1}^{k}\frac{\sigma_{y,j}^2(1-\rho_j^2\alpha_{x,j}\alpha_{y,j})}{\alpha_{y,j}}H_j\sum_{j=1}^{k-1}\omega_jG_j \leq 0\right]$$

where k is the number of moderator-based subpopulations, $\sigma_{y,j}^2$ is the variance of the true Y scores in subpopulation j, ρ_j^2 is the correlation between the true X and Y scores in subpopulation j, $\alpha_{x,j}$ is the reliability for X in subpopulation j, $\alpha_{y,j}$ is the reliability for Y in subpopulation j, and ω_j is the jth eigenvalue of $(\mathbf{C'DC})^{-1}\mathbf{C'VC}$:

$$\mathbf{D} = \text{Diag}\left[\frac{\alpha_{x,j}(n_j+1)}{(n_j-1)^2\delta_j\sigma_{x,j}^2};j=1,...,k\right]$$

$$\mathbf{V} = \text{Diag}\left[\frac{\sigma_{y,j}^2\alpha_{x,j}(1-\rho_j^2\alpha_{x,j}\alpha_{y,j})(n_j+1)}{\alpha_{y,j}(n_j-1)^2\delta_j\sigma_{x,j}^2};j=1,...,k\right]$$

where n_j is the size of subpopulation j, δ_j is the ratio of the expected sample variance of X to the population variance of X in subpopulation j, $\sigma_{x,j}^2$ is the variance of the true X scores in subpopulation j, and G_j for $j = 1, \ldots, k-1$ and H_j for $j = 1, \ldots, k$ are independently distributed chi-squared random variables. Specifically, $H_j \sim \chi^2(n_j - 2)$ for

From Aguinis, Boik, and Pierce (2001).

173

$j = 1, \ldots, k$ and $G_j \sim \chi^2(1, \lambda_j)$ for $j = 1, \ldots, k - 1$, where λ_j is a noncentrality parameter:

$$\lambda_j = \frac{(\mathbf{u}_j' \mathbf{C}' \boldsymbol{\beta}_1)^2}{2\mathbf{u}_j' \mathbf{C}' \mathbf{V} \mathbf{C} \mathbf{u}_j}$$

and $\mathbf{u}_j$ is the jth eigenvector of $(\mathbf{C}'\mathbf{D}\mathbf{C})^{-1} \mathbf{C}'\mathbf{V}\mathbf{C}$.

References

Abdel-Halim, A. A. (1980). Effects of higher order need strength on the job per-
    formance-job satisfaction relationship. *Personnel Psychology, 33,* 335–248.
Abelson, R. P. (1952). Sex differences in predictability of college grades. *Educa-
    tional and Psychological Measurement, 12,* 638–644.
Aguinis, H. (1993). Action research and scientific method: Presumed discrep-
    ancies and actual similarities. *Journal of Applied Behavioral Science, 29,*
    416–431.
Aguinis, H. (1994). A QuickBASIC program for generating correlated multi-
    variate random normal scores. *Educational and Psychological Measurement,
    54,* 687–689.
Aguinis, H. (1995). Statistical power problems with moderated multiple regres-
    sion in management research. *Journal of Management, 21,* 1141–1158.
Aguinis, H. (2002). Estimation of interaction effects in organization studies.
    Organizational Research Methods, 5, 207–211.
Aguinis, H., & Adams, S. K. R. (1998). Social-role versus structural models of
    gender and influence use in organizations: A strong inference approach.
    Group and Organization Management, 23, 414–446.
Aguinis, H., Beaty, J. C., Boik, R. J., & Pierce, C. A. (2003). *Assessing moderating
    effects of categorical variables using multiple regression: A 30-year review.*
    Manuscript submitted for publication.
Aguinis, H., Boik, R. J., & Pierce, C. A. (2001). A generalized solution for ap-
    proximating the power to detect effects of categorical moderator variables
    using multiple regression. *Organizational Research Methods, 4,* 291–323.
Aguinis, H., Bommer, W. H., & Pierce, C. A. (1996). Improving the estimation
    of moderating effects by using computer-administered questionnaires. *Ed-
    ucational and Psychological Measurement, 56,* 1043–1047.
Aguinis, H., & Henle, C. A. (2002). Ethics in research. In S. G. Rogelberg (Ed.),
    Handbook of research methods in industrial and organizational psychology
    (pp. 34–56). Malden, MA: Blackwell.
Aguinis, H., Henle, C. A., & Ostroff, C. (2001). Measurement in work and or-
    ganizational psychology. In N. Anderson, D. S. Ones, H. K. Sinangil, &
    C. Viswesvaran (Eds.), *Handbook of industrial, work and organizational psy-
    chology* (Vol. 1, pp. 27–50). London: Sage.

Aguinis, H., Nesler, M. S., Quigley, B. M., Lee, S., & Tedeschi, J. T. (1996). Power bases of faculty supervisors and educational outcomes for graduate students. *Journal of Higher Education, 67,* 267–297.

Aguinis, H., Petersen, S. A., & Pierce, C. A. (1999). Appraisal of the homogeneity of error variance assumption and alternatives to multiple regression for estimating moderating effects of categorical variables. *Organizational Research Methods, 2,* 315–339.

Aguinis, H., & Pierce, C. A. (1998a). Heterogeneity of error variance and the assessment of moderating effects of categorical variables: A conceptual review. *Organizational Research Methods, 1,* 296–314.

Aguinis, H., & Pierce, C. A. (1998b). Statistical power computations for detecting dichotomous moderator variables with moderated multiple regression. *Educational and Psychological Measurement, 58,* 668–676.

Aguinis, H., & Pierce, C. A. (1998c). Testing moderator variable hypotheses meta-analytically. *Journal of Management, 24,* 577–592.

Aguinis, H., Pierce, C. A., & Quigley, B. M. (1993). Conditions under which a bogus pipeline procedure enhances the validity of self-reported cigarette smoking: A meta-analytic review. *Journal of Applied Social Psychology, 23,* 352–373.

Aguinis, H., Pierce, C. A., & Stone-Romero, E. F. (1994). Estimating the power to detect dichotomous moderators with moderated multiple regression. *Educational and Psychological Measurement, 54,* 690–692.

Aguinis, H., & Roth, H. A. (2003). *Teaching in China: Culture-based challenges.* Manuscript submitted for publication.

Aguinis, H., & Stone-Romero, E. F. (1997). Methodological artifacts in moderated multiple regression and their effects on statistical power. *Journal of Applied Psychology, 82,* 192–206.

Aguinis, H., & Whitehead, R. (1997). Sampling variance in the correlation coefficient under indirect range restriction: Implications for validity generalization. *Journal of Applied Psychology, 82,* 528–538.

Aiken, L. S., & West, S. G. (1991). *Multiple regression: Testing and interpreting interactions.* Newbury Park, CA: Sage.

Aiken, L. S., & West, S. G. (1993). Detecting interactions in multiple regression: Measurement error, power, and design considerations. *The Score, 16*(1), 7, 14–15.

Alexander, R. A., Barrett, G. V., Alliger, G. M., & Carson, K. P. (1986). Toward a general model of nonrandom sampling and the impact on population correlations: Generalizations of Berkson's Fallacy and restriction of range. *British Journal of Mathematical and Statistical Psychology, 39,* 90–105.

Alexander, R. A., & DeShon, R. P. (1994). Effect of error variance heterogeneity on the power of tests for regression slope differences. *Psychological Bulletin, 115,* 308–314.

Alexander, R. A., & Govern, D. M. (1994). A new and simpler approximation for ANOVA under variance heterogeneity. *Journal of Educational Statistics, 19,* 91–101.

American Educational Research Association, American Psychological Association, & National Council on Measurement in Education. (1999). *Standards for educational and psychological testing.* Washington, DC: Author.

Anderson, L. E., Stone-Romero, E. F., & Tisak, J. A. (1996). A comparison of bias and mean squared error in parameter estimates of interaction effects: Moderated multiple regression versus errors-in-variables regression. *Multivariate Behavioral Research, 31,* 69–94.

Arnold, H. J. (1981). A test of the multiplicative hypothesis of expectancy valence theories of work motivation. *Academy of Management Journal, 24,* 128–141.

Arnold, H. J. (1982). Moderator variables: A clarification of conceptual, analytic, and psychometric issues. *Organizational Behavior and Human Performance, 29,* 143–174.

Arnold, H. J. (1984). Testing moderator variable hypotheses: A reply to Stone and Hollenbeck. *Organizational Behavior and Human Performance, 34,* 214–224.

Arnold, H. J., & Evans, M. G. (1979). Testing multiplicative models does not require ratio scales. *Organizational Behavior and Human Performance, 24,* 41–59.

Association to Advance Collegiate Schools of Business. (2001). *2000–2001 salary survey results.* Available online at *http://www.aacsb.edu/Publications/ Newsline/view.asp?year=2001&file=wnsalsurv_1.html.*

Avolio, B. J., Waldman, D. A., & McDaniel, M. A. (1990). Age and work performance in nonmanagerial jobs: The effects of experience and occupation type. *Academy of Management Journal, 33,* 407–422.

Baker, J. R., & Yardley, J. K. (2002). Moderating effect of gender on the relationship between sensation seeking-impulsivity and substance use in adolescents. *Journal of Child and Adolescent Substance Abuse, 12,* 27–43.

Baron, R. M., & Kenny, D. A. (1986). The moderator–mediator variable distinction in social psychological research: Conceptual, strategic, and statistical considerations. *Journal of Personality and Social Psychology, 51,* 1173–1182.

Bartlett, C. J., Bobko, P., Mosier, S. B., & Hannan, R. (1978). Testing for fairness with a moderated regression strategy: An alternative to differential analysis. *Personnel Psychology, 31,* 223–241.

Bartlett, C. J., & O'Leary, B. S. (1969). A differential prediction model to moderate the effects of heterogeneous groups in personnel selection and classification. *Personnel Psychology, 22,* 1–17.

Bartlett, M. S. (1937). Properties of sufficiency and statistical tests. *Proceedings of the Royal Society, A160,* 268–282.

Baumeister, R. F. (1990). Item variances and median splits: Some encouraging and reassuring findings. *Journal of Personality, 58,* 589–594.

Bedeian, A. G., & Mossholder, K. W. (1994). Simple question, not so simple answer: Interpreting interaction terms in moderated multiple regression. *Journal of Management, 20,* 159–165.

Bhagat, R. S. (1982). Conditions under which stronger job performance–job satisfaction relationships may be observed: A closer look at two situational contingencies. *Academy of Management Journal, 25,* 772–789.

Bissonnette, V., Ickes, W., Bernstein, I., & Knowles, E. (1990). Personality moderating variables: A warning about statistical artifact and a comparison of analytic techniques. *Journal of Personality, 58,* 567–587.

Blood, M. R., & Mullet, G. M. (1977). *Where have all the moderators gone: The perils of Type II error.* Atlanta: College of Industrial Management, Georgia Institute of Technology.

Bobko, P. (1986). A solution to some dilemmas when testing hypotheses about ordinal interactions. *Journal of Applied Psychology, 71,* 323–326.

Bobko, P., & Bartlett, C. J. (1978). Subgroup validities: Differential definitions and differential prediction. *Journal of Applied Psychology, 63,* 12–14.

Bobko, P., & Russell, C. J. (1994). On theory, statistics, and the search for interactions in the organizational sciences. *Journal of Management, 20,* 193–200.

Boehm, V. R. (1977). Differential prediction: A methodological artifact? *Journal of Applied Psychology, 62,* 146–154.

Bohrnstedt, G. W., & Marwell, G. (1978). The reliability of products of two random variables. In K. F. Schuessler (Ed.), *Sociological methodology* (pp. 254–273). San Francisco: Jossey-Bass.

Boik, R. J. (1979). Interactions, partial interactions, and interaction contrasts in the analysis of variance. *Psychological Bulletin, 86,* 1084–1089.

Boik, R. J. (1993). The analysis of two-factor interactions in fixed effects linear models. *Journal of Educational Statistics, 18,* 1–40.

Busemeyer, J. R., & Jones, L. E. (1983). Analysis of multiplicative combination rules when the causal variables are measured with error. *Psychological Bulletin, 93,* 549–562.

Cascio, W. F., & Aguinis, H. (2001). The Federal Uniform Guidelines on Employee Selection Procedures (1978): An update on selected issues. *Review of Public Personnel Administration, 21,* 200–218.

Cascio, W. F., & Aguinis, H. (2003). *Test development and use: New twists on old questions.* Manuscript submitted for publication.

Cascio, W. F., & Zedeck, S. (1983). Open a new window in rational research planning: Adjust alpha to maximize statistical power. *Personnel Psychology, 36,* 517–526.

Casella, G., & Berger, R. L. (2002). *Statistical inference* (2nd ed.). Belmont, CA: Duxbury Press.

Champoux, J. E., & Peters, W. S. (1980). Applications of moderated regression in job design research. *Personnel Psychology, 33,* 759–783.

Champoux, J. E., & Peters, W. S. (1987). Form, effect size and power in moderated regression analysis. *Journal of Occupational Psychology, 60,* 243–255.

Chaplin, W. F. (1997). Personality, interactive relations, and applied psychology. In R. Hogan, J. Johnson, & S. Briggs (Eds.), *Handbook of personality psychology* (pp. 873–890). San Diego, CA: Academic Press.

Chaplin, W. F., & Goldberg, L. R. (1984). A failure to replicate the Bem and Allen study on individual differences in cross-situational consistencies. *Journal of Personality and Social Psychology, 47,* 1074–1090.

Cheek, J. M. (1989). Identity orientations and self-interpretation. In D. M. Buss & N. Cantor (Eds.), *Personality psychology, recent trends and emerging directions* (pp. 275–285). New York: Springer-Verlag.

Chow, S. L. (1996). *Statistical significance: Rationale, validity and utility.* Thousand Oaks, CA: Sage.

Cicchetti, D. V., Showalter, D., & Tyrer, P. J. (1985). The effect of number of rating scale categories on levels of interrater reliability: A Monte Carlo investigation. *Applied Psychological Measurement, 9,* 31–36.

Cleary, T. A. (1968). Test bias: Prediction of grades of Negro and white students in integrated colleges. *Journal of Educational Measurement, 5,* 115–124.

Cohen, J. (1962). The statistical power of abnormal-social psychological research: A review. *Journal of Abnormal and Social Psychology, 65,* 145–153.

Cohen, J. (1977). *Statistical power analysis for the behavioral sciences* (rev. ed.). New York: Academic Press.

Cohen, J. (1978). Partialed products are interactions; partialed powers are curve components. *Psychological Bulletin, 85,* 858–866.

Cohen, J. (1983). The cost of dichotomization. *Applied Psychological Measurement, 7,* 249–253.

Cohen, J. (1988). *Statistical power analysis for the behavioral sciences* (2nd ed.). Hillsdale, NJ: Erlbaum.

Cohen, J. (1994). The earth is round ($p < 0.05$). *American Psychologist, 49,* 997–1003.

Cohen, J., & Cohen, P. (1983). *Applied multiple regression/correlational analysis for the behavioral sciences* (2nd ed.). Hillsdale, NJ: Erlbaum.

Cohen, J., Cohen, P., West, S. G., & Aiken, L. S. (2003). *Applied multiple regression/correlation analysis for the behavioral sciences* (3rd ed.). Mahwah, NJ: Erlbaum.

Conway, D., & Roberts, H. V. (1983). Reverse regression, fairness, and employment discrimination. *Journal of Business and Economic Statistics, 1,* 75–85.

Conyon, M. J., & Peck, S. I. (1998). Board control, remuneration committees, and top management compensation. *Academy of Management Journal, 41,* 146–157.

Cortina, J. M. (1993). Interaction, nonlinearity, and multicollinearity: Implications for multiple regression. *Journal of Management, 19,* 915–922.

Cortina, J. M., & DeShon, R. P. (1998). Determining relative importance of predictors with the observational design. *Journal of Applied Psychology, 83,* 798–804.

Cortina, J. M., & Dunlap, W. P. (1997). On the logic and purpose of significance testing. *Psychological Methods, 2,* 161–172.

Cortina, J. M., & Folger, R. G. (1998). When is it acceptable to accept a null hypothesis: No way, Jose? *Organizational Research Methods, 1,* 334–350.

Court, A. T. (1930). Measuring joint causation. *Journal of the American Statistical Association, 25,* 245–254.

Cronbach, L. J. (1987). Statistical tests for moderator variables: Flaws in analyses recently proposed. *Psychological Bulletin, 102,* 414–417.

Dance, K. A., & Neufeld, R. W. J. (1988). Aptitude-treatment interaction research in the clinical setting: A review of attempts to dispel the "patient uniformity" myth. *Psychological Bulletin, 104,* 192–213.

Darrow, A. L., & Kahl, D. R. (1982). A comparison of moderated regression techniques considering strength of effect. *Journal of Management, 8,* 35–47.

DeShon, R. P., & Alexander, R. A. (1994). A generalization of James's second-order approximation to the test for regression slope equality. *Educational and Psychological Measurement, 54,* 328–335.

DeShon, R. P., & Alexander, R. A. (1996). Alternative procedures for testing regression slope homogeneity when group error variances are unequal. *Psychological Methods, 1,* 261–277.

Diaconis, P., & Efron, B. (1983). Computer-intensive methods in statistics. *Scientific American, 248*(5), 116–130.

Drasgow, F., & Kang, T. (1984). Statistical power of differential validity and differential prediction analyses for detecting measurement nonequivalence. *Journal of Applied Psychology, 69,* 498–508.

Dretzke, B. J., Levin, J. R., & Serlin, R. C. (1982). Testing for regression homogeneity under variance heterogeneity. *Psychological Bulletin, 91,* 376–383.

Dunlap, W. P., & Kemery, E. R. (1987). Failure to detect moderating effects: Is multicollinearity the problem? *Psychological Bulletin, 102,* 418–420.

Dunlap, W. P., & Kemery, E. R. (1988). Effects of predictor intercorrelations and reliabilities on moderated multiple regression. *Organizational Behavior and Human Decision Processes, 41,* 248–258.

Eden, D. (2002). Replication, meta-analysis, scientific progress, and AMJ's publication policy. *Academy of Management Journal, 45,* 841–846.

Edwards, W. (1954). The theory of decision making. *Psychological Bulletin, 51,* 380–417.

Efron, B. (1979). Bootstrap methods: Another look at the jackknife. *Annals of Statistics, 7,* 1–26.

Efron, B., & Tibshirani, R. J. (1993). *An introduction to the bootstrap.* New York: Chapman & Hall.

Einhorn, H. J., & Bass, A. R. (1971). Methodological considerations relevant to discrimination in employment testing. *Psychological Bulletin, 75,* 261–269.

Evans, M. G. (1985). A Monte Carlo study of the effects of correlated method variance in moderated multiple regression analysis. *Organizational Behavior and Human Decision Processes, 36,* 305–323.

Evans, M. G. (1991a). On the use of moderated regression. *Canadian Psychology, 32,* 116–119.

Evans, M. G. (1991b). The problem of analyzing multiplicative composites: Interactions revisited. *American Psychologist, 46,* 6–15.

Fisicaro, S. A., & Lautenschlager, G. J. (1992). Power and reliability: The case of homogeneous true score regression across treatments. *Educational and Psychological Measurement, 52,* 505–511.

Fisicaro, S. A., & Tisak, J. (1994). A theoretical note on the stochastics of moderated multiple regression. *Educational and Psychological Measurement, 54,* 32–41.

Flanagan, J. C., Dailey, J. T., Shaycoft, M. F., Gorham, W. A., Orr, D. B., & Goldberg, I. (1962). *Design for a study of American youth.* Boston: Houghton Mifflin.

Frederiksen, N., & Melville, S. D. (1954). Differential predictability in the use of test scores. *Educational and Psychological Measurement, 14,* 647–656.

Friedrich, R. J. (1982). In defense of multiplicative terms in multiple regression equations. *American Journal of Political Science, 26,* 797–833.

Frone, M. R. (1999). Work stress and alcohol use. *Alcohol Research and Health, 23,* 284–291.

Frone, M. R., Russell, M. R., & Cooper, M. L. (1997). Job stressors, job involvement, and employee health: A test of identity theory. *Journal of Occupational and Organizational Psychology, 68*, 1–11.

Fry, L. W., Kerr, S., & Lee, C. (1986). Effects of different leader behaviors under different levels of task interdependence. *Human Relations, 39*, 1067–1082.

Games, P. A., Winkler, H. B., & Probert, D. A. (1972). Robust tests for homogeneity of variance. *Educational and Psychological Measurement, 32*, 887–909.

Ganzach, Y. (1997). Misleading interaction and curvilinear terms. *Psychological Methods, 2*, 235–247.

Ganzach, Y. (1998). Nonlinearity, multicollinearity and the probability of Type II error in detecting interaction. *Journal of Management, 24*, 615–622.

Ganzach, Y., Saporta, I., & Weber, Y. (2000). Interactions in linear versus logistic models: A substantive illustration using the relationship between motivation, ability, and performance. *Organizational Research Methods, 3*, 237–253.

Gartside, P. A. (1972). A study of methods for comparing several variances. *Journal of the American Statistical Association, 67*, 342–346.

Gaylord, R. H., & Carroll, J. B. (1948). A general approach to the problem of the population control variable. *American Psychologist, 3*, 310.

Gerard, P. D., Smith, D. R., & Weerakkody, G. (1998). Limits of retrospective power analysis. *Journal of Wildlife Management, 62*, 801–807.

Ghiselli, E. E. (1956). Differentiation of individuals in terms of their predictability. *Journal of Applied Psychology, 40*, 374–377.

Ghiselli, E. E. (1960a). Differentiation of tests in terms of the accuracy with which they predict a given individual. *Educational and Psychological Measurement, 20*, 675–684.

Ghiselli, E. E. (1960b). The prediction of predictability. *Educational and Psychological Measurement, 20*, 3–8.

Goodman, S. N., & Berlin, J. A. (1994). The use of predicted confidence intervals when planning experiments and the misuse of power when interpreting results. *Annals of Internal Medicine, 121*, 200–206.

Greenberger, E., Chen, C., Tally, S. R., & Dong, Q. (2000). Family, peer, and individual correlates of depressive symptomatology among U.S. and Chinese adolescents. *Journal of Consulting and Clinical Psychology, 68*, 209–219.

Grooms, R. R., & Endler, N. S. (1960). The effect of anxiety on academic achievement. *Journal of Educational Psychology, 51*, 299–304.

Gulliksen, H., & Wilks, S. S. (1950). Regression tests for several samples. *Psychometrika, 15*, 91–114.

Hackman, J. R., & Oldham, G. R. (1976). Motivation through the design of work: Test of a theory. *Organizational Behavior and Human Performance, 16*, 250–279.

Hall, J. A., & Rosenthal, R. (1991). Testing for moderator variables in meta-analysis: Issues and methods. *Communication Monographs, 58*, 437–448.

Hartmann, F. G. H., & Moers, F. (1999). Testing contingency hypotheses in budgetary research: An evaluation of the use of moderated regression analysis. *Accounting, Organizations, and Society, 24*, 291–315.

Hoenig, J. M., & Heisey, D. M. (2001). The abuse of power: The pervasive fallacy of power calculations for data analysis. *American Statistician, 55*, 19–24.

Hsu, L. M. (1993). Using Cohen's tables to determine the maximum power attainable in two-sample tests when one sample is limited in size. *Journal of Applied Psychology, 78,* 303–305.

Huff, D. (1954). *How to lie with statistics.* New York: Norton.

Hunter, J. E., & Schmidt, F. L. (1978). Differential and single-group validity of employment tests by race: A critical analysis of three recent studies. *Journal of Applied Psychology, 63,* 1–11.

Hunter, J. E., & Schmidt, F. L. (1990). *Methods of meta-analysis: Correcting error and bias in research findings.* Newbury Park, CA: Sage.

Hunter, J. E., Schmidt, F. L., & Hunter, R. (1979). Differential validity of employment tests by race: A comprehensive review and analysis. *Psychological Bulletin, 86,* 721–735.

Hunter, J. E., Schmidt, F. L., & Le, H. (2002). *Implications of direct and indirect range restriction for meta-analysis methods and findings.* Unpublished manuscript, Department of Psychology, Michigan State University.

Hunter, J. E., Schmidt, F. L., & Rauschenberger, J. (1984). Methodological and statistical issues in the study of bias in mental testing. In C. R. Reynolds & R. T. Brown (Eds.), *Perspectives on bias in mental testing* (pp. 41–100). New York: Plenum.

Irwin, J. R., & McClelland, G. H. (2001). Misleading heuristics and moderated multiple regression models. *Journal of Marketing Research, 38,* 100–109.

Jaccard, J. J. (2001). *Interaction effects in logistic regression.* Newbury Park, CA: Sage.

Jaccard, J. J., Turrisi, R., & Wan, C. K. (1990). *Interaction effects in multiple regression.* Newbury Park, CA: Sage.

Jaccard, J., & Wan, C. K. (1995). Measurement error in the analysis of interaction effects between continuous predictors using multiple regression: Multiple indicator and structural equation approaches. *Psychological Bulletin, 117,* 348–357.

James, G. S. (1951). The comparison of several groups of observations when the ratios of the population variances are unknown. *Biometrika, 38,* 324–329.

James, L. R., & Brett, J. M. (1984). Mediators, moderators, and tests for mediation. *Journal of Applied Psychology, 69,* 307–321.

Johnson, D. C. (1960). The population control variable or moderator variable in personnel research. In *Tri-Service Conference on Selection Research* (pp. 125–134). Washington, DC: Office of Naval Research.

Johnson, D. H. (2001). Sharing data: It's time to end psychology's guild approach. *Observer, 14*(8), 1, 38–39.

Jones, M. B. (1973). Moderated regression and equal opportunity. *Educational and Psychological Measurement, 33,* 591–602.

Jorm, A. F., Christensen, H., Henderson, A. S., Jacomb, P. A., Korten, A. E., & Rodgers, B. (2000). Predicting anxiety and depression from personality: Is there a synergistic effect of neuroticism and extraversion? *Journal of Abnormal Psychology, 109,* 145–149.

Judd, C. M., Kenny, D. A., & McClelland, G. H. (2001). Estimating and testing mediation and moderation in within-subjects designs. *Psychological Methods, 6,* 115–134.

Kahl, D. R., & Darrow, A. L. (1984). Model determination in moderated regression. *Journal of Management, 10*, 234–236.

Kanetkar, V., Evans, M. G., Everell, S. A., Irvine, D., & Millman, Z. (1995). The effect of scale changes on meta-analysis of multiplicative and main effects models. *Educational and Psychological Measurement, 55*, 206–224.

Katrichis, J. (1993). The conceptual implications of data centering in interactive regression models. *Journal of the Market Research Society, 35*, 183–192.

Keppel, G., & Zedeck, S. (1989). *Data analysis for research designs: Analysis of variance and multiple regression/correlation approaches.* New York: Freeman.

Kerr, S. (1975). On the folly of rewarding A, while hoping for B. *Academy of Management Journal, 18*, 769–783.

Kowalski, R. M. (1995). Teaching moderated multiple regression for the analysis of mixed factorial designs. *Teaching of Psychology, 22*, 197–198.

Kromrey, J. D., & Foster-Johnson, L. (1998). Mean centering in moderated multiple regression: Much ado about nothing. *Educational and Psychological Measurement, 58*, 42–67.

Kromrey, J. D., & Foster-Johnson, L. (1999). Statistically differentiating between interaction and nonlinearity in multiple regression analysis: A Monte Carlo investigation of a recommended strategy. *Educational and Psychological Measurement, 59*, 392–413.

LaCour-Little, M. (1996). Application of reverse regression to Boston Federal Reserve data refutes claims of discrimination. *Journal of Real Estate Research, 11*, 1–12.

Landis, R. S., & Dunlap, W. P. (2000). Moderated multiple regression tests are criterion specific. *Organizational Research Methods, 3*, 254–266.

Lautenschlager, G. J., & Mendoza, J. L. (1986). A step-down hierarchical multiple regression analysis for examining hypotheses about test bias in prediction. *Applied Psychological Measurement, 10*, 133–139.

LeClerc, F., & Little, J. D. C. (1997). Can advertising copy make FSI coupons more effective? *Journal of Marketing Research, 34*, 473–484.

Linn, R. L. (1968). Range restriction problems in the use of self-selected groups for test validation. *Psychological Bulletin, 69*, 69–73.

Lubinski, D., & Humphreys, L. G. (1990). Assessing spurious "moderator effects": Illustrated substantively with the hypothesized ("synergistic") relation between spatial and mathematical ability. *Psychological Bulletin, 107*, 383–393.

MacCallum, R. C., & Mar, C. M. (1995). Distinguishing between moderator and quadratic effects in multiple regression. *Psychological Bulletin, 118*, 405–421.

Marascuilo, L. A. (1966). Large-sample multiple comparisons. *Psychological Bulletin, 65*, 280–290.

Markus, K. A. (2001). The converse inequality argument against tests of statistical significance. *Psychological Methods, 6*, 147–160.

Marshall, R. S., & Boush, D. M. (2001). Dynamic decision-making: A cross-cultural comparison of U.S. and Peruvian export managers. *Journal of International Business Studies, 32*, 873–893.

Mason, C. A., Tu, S., & Cauce, A. M. (1996). Assessing moderator variables:

Two computer simulation studies. *Educational and Psychological Measurement, 56,* 45–62.

Maxwell, S. E., & Delaney, H. D. (1993). Bivariate median splits and spurious statistical significance. *Psychological Bulletin, 113,* 181–190.

Mazen, A. M., Graf, L. A., Kellogg, C. E., & Hemmasi, M. (1987). Statistical power in contemporary management research. *Academy of Management Journal, 30,* 369–380.

Mazen, A. M., Hemmasi, M., & Lewis, M. F. (1987). Assessment of statistical power in contemporary strategy research. *Strategic Management Journal, 8,* 403–410.

McClelland, G. H., & Judd, C. M. (1993). Statistical difficulties of detecting interactions and moderator effects. *Psychological Bulletin, 114,* 376–390.

Mone, M. A., Mueller, G. C., & Mauland, W. (1996). The perceptions and usage of statistical power in applied psychology and management research. *Personnel Psychology, 49,* 103–120.

Morris, J. H., Sherman, J., & Mansfield, E. R. (1986). Failures to detect moderating effects with ordinary least squares moderated-regression: Some reasons and a remedy. *Psychological Bulletin, 99,* 282–288.

Mossholder, K. W., Kemery, E. R., & Bedeian, A. G. (1990). On using regression coefficients to interpret moderator effects. *Educational and Psychological Measurement, 50,* 255–263.

Murphy, K. R. (2002). Using power analysis to evaluate and improve research. In S. G. Rogelberg (Ed.), *Handbook of research methods in industrial and organizational psychology* (pp. 119–137). Malden, MA: Blackwell.

Murphy, K., & Myors, B. (1998). *Statistical power analysis: A simple and general model for traditional and modern hypothesis tests.* Mahwah, NJ: Erlbaum.

Nickerson, R. S. (2000). Null hypothesis significance testing: A review of an old and continuing controversy. *Psychological Methods, 5,* 241–301.

O'Connor, B. P. (1998). SIMPLE: All-in-one programs for exploring interactions in moderated multiple regression. *Educational and Psychological Measurement, 58,* 836–840.

Oswald, F. L., Saad, S., & Sackett, P. R. (2000). The homogeneity assumption in differential prediction analysis: Does it really matter? *Journal of Applied Psychology, 85,* 536–541.

Overall, J. E., Lee, D. M., & Hornick, C. W. (1981). Comparisons of two strategies for analyses of variance in nonorthogonal decisions. *Psychological Bulletin, 90,* 367–375.

Overton, R. C. (2001). Moderated multiple regression for interactions involving categorical variables: A statistical control for heterogeneous variance across two groups. *Psychological Methods, 6,* 218–233.

Paunonen, S. V., & Jackson, D. N. (1988). Type I error rates for moderated multiple regression analysis. *Journal of Applied Psychology, 73,* 569–573.

Pedhazur, E. J. (1982). *Multiple regression in behavioral research.* Orlando, FL: Holt, Rinehart & Winston.

Raju, N. S., Fralicx, R., & Steinhaus, S. D. (1986). Covariance and regression slope models for studying validity generalization. *Applied Psychological Measurement, 10,* 195–211.

Raju, N. S., Pappas, S., & Williams, C. P. (1989). An empirical Monte Carlo test of the accuracy of the correlation, covariance, and regression slope models for assessing validity generalization. *Journal of Applied Psychology, 74,* 901–911.

Rasmussen, J. L. (1989). A Monte Carlo evaluation of Bobko's ordinal interaction analysis technique. *Journal of Applied Psychology, 74,* 242–246.

Rock, D. A., Barone, J. L., & Linn, R. L. (1967). A FORTRAN computer program for a moderated stepwise prediction system. *Educational and Psychological Measurement, 27,* 709–713.

Rogers, W. M. (2002). Theoretical and mathematical constraints of interaction regression models. *Organizational Research Methods, 5,* 212–230.

Rogosa, D. (1980). Comparing nonparallel regression lines. *Psychological Bulletin, 88,* 307–321.

Rosenthal, R. (1979). The "file drawer problem" and tolerance for null results. *Psychological Bulletin, 86,* 638–641.

Rosenthal, R., & Rosnow, R. L. (1985). *Contrast analysis: Focused comparisons in the analysis of variance.* Cambridge, UK: Cambridge University Press.

Rosenthal, R., Rosnow, R. L., & Rubin, D. B. (2000). *Contrasts and effect sizes in behavioral research: A correlational approach.* Cambridge, UK: Cambridge University Press.

Rotundo, M., & Sackett, P. R. (1999). Effect of rater race on conclusions regarding differential prediction in cognitive ability tests. *Journal of Applied Psychology, 84,* 815–822.

Russell, C. J., & Bobko, P. (1992). Moderated regression analysis and Likert scales: Too coarse for comfort. *Journal of Applied Psychology, 77,* 336–342.

Russell, C. J., & Dean, M. A. (2000). To log or not to log: Bootstrap as an alternative to parametric estimation of moderation effects in the presence of skewed dependent variables. *Organizational Research Methods, 3,* 167–185.

Russell, C. J., Pinto, J., & Bobko, P. (1991). Appropriate moderated regression and inappropriate research strategy: A demonstration of the need to give your respondents space. *Applied Psychological Measurement, 15,* 257–266.

Russell, C. J., Settoon, R. P., McGrath, R. N., Blanton, A. E., Kidwell, R. E., Lohrke, F. T., Scifres, E. L., & Danforth, G. W. (1994). Investigator characteristics as moderators of personnel selection research: A meta-analysis. *Journal of Applied Psychology, 79,* 163–170.

Saad, S., & Sackett, P. R. (2002). Investigating differential prediction by gender in employment-oriented personality measures. *Journal of Applied Psychology, 87,* 667–674.

Sackett, P. R., & Wilk, S. L. (1994). Within-group norming and other forms of source adjustment in preemployment testing. *American Psychologist, 49,* 929–954.

Salgado, J. F. (1998). Sample size in validity studies of personnel selection. *Journal of Occupational and Organizational Psychology, 71,* 161–164.

Saunders, D. R. (1955). The "moderator variable" as a useful tool in prediction. In *Proceedings of the 1954 Invitational Conference on Testing Problems* (pp. 54–58). Princeton, NJ: Educational Testing Service.

Saunders, D. R. (1956). Moderator variables in prediction. *Educational and Psychological Measurement, 16,* 209–222.

Schmidt, F. L. (1996). Statistical significance testing and cumulative knowledge in psychology: Implications for training of researchers. *Psychological Methods, 1,* 115–129.

Schmidt, F. L. (2002). The role of general cognitive ability in job performance: Why there cannot be a debate. *Human Performance, 15,* 187–210.

Schoorman, F. D., Bobko, P., & Rentsch, J. (1991). The role of theory in testing hypothesized interactions: An example from the research on escalation of commitment. *Journal of Applied Social Psychology, 21,* 1338–1355.

Sechrest, L. (2000, June). *Statistical power: Its uses and abuses.* Methodology tutorial presented at the meeting of the American Psychological Society, Miami Beach, FL.

Sedlmeier, P., & Gigerenzer, G. (1989). Do studies of statistical power have an effect on the power of studies? *Psychological Bulletin, 105,* 309–316.

Serlin, R. C., & Levin, J. R. (1985). Teaching how to derive directly interpretable coding schemes for multiple regression. *Journal of Educational Statistics, 10,* 223–238.

Sharma, S., Durand, R. M., & Gur-Arie, O. (1981). Identification and analysis of moderator variables. *Journal of Marketing Research, 18,* 291–300.

Sheeran, P., & Abraham, C. (2003). Mediator of moderators: Temporal stability on intention and the intention-behavior relation. *Personality and Social Psychology Bulletin, 29,* 205–215.

Shepperd, J. A. (1991). Cautions in assessing spurious "moderator effects." *Psychological Bulletin, 110,* 315–317.

Shrout, P. E., & Bolger, N. (2002). Mediation in experimental and nonexperimental studies: New procedures and recommendations. *Psychological Methods, 7,* 422–445.

Smith, B., & Sechrest, L. (1991). Treatment of aptitude × treatment interactions. *Journal of Consulting and Clinical Psychology, 59,* 233–244.

Smith, K. W., & Sasaki, M. S. (1979). Decreasing multicollinearity: A method for models with multiplicative functions. *Sociological Methods and Research, 8,* 35–56.

Society for Industrial and Organizational Psychology, Inc. (1987). *Principles for the validation and use of personnel selection procedures* (3rd ed.). College Park, MD: Author.

SPSS, Inc. (1999). *SPSS base 10.0 applications guide.* Chicago: Author.

Starbuck, B., & Mezias, J. (1996). Journal impact ratings. *The Industrial-Organizational Psychologist, 33*(4), 101–105.

Stewart, G. L., Carson, K. P., & Cardy, R. L. (1996). The joint effects of conscientiousness and self-leadership training on employee self-directed behavior in a service setting. *Personnel Psychology, 49,* 143–164.

St. John, C. H., & Roth, P. L. (1999). The impact of cross-validation adjustments on estimates of effect size in business policy and strategy research. *Organizational Research Methods, 2,* 157–174.

Stone, E. F. (1986). Research methods in industrial and organizational psychology: Selected issues and trends. In C. L. Cooper & I. T. Robertson (Eds.), *International review of industrial and organizational psychology* (pp. 305–334). London: Wiley.

Stone, E. F. (1988). Moderator variables in research: A review and analysis of conceptual and methodological issues. In G. R. Ferris & K. M. Rowland

(Eds.), *Research in personnel and human resources management* (Vol. 6, pp. 191–229). Greenwich, CT: JAI Press.

Stone, E. F., & Hollenbeck, J. R. (1984). Some issues associated with the use of moderated regression. *Organizational Behavior and Human Performance, 34,* 195–213.

Stone, E. F., & Hollenbeck, J. R. (1989). Clarifying some controversial issues surrounding statistical procedures for detecting moderator variables: Empirical evidence and related matters. *Journal of Applied Psychology, 74,* 3–10.

Stone-Romero, E. F., Alliger, G. M., & Aguinis, H. (1994). Type II error problems in the use of moderated multiple regression for the detection of moderating effects of dichotomous variables. *Journal of Management, 20,* 167–178.

Stone-Romero, E. F., & Anderson, L. E. (1994). Techniques for detecting moderating effects: Relative statistical power of multiple regression and the comparison of subgroup-based correlation coefficients. *Journal of Applied Psychology, 79,* 354–359.

Stone-Romero, E. F., & Liakhovitski, D. (2002). Strategies for detecting moderator variables: A review of conceptual and empirical issues. *Research in Personnel and Human Resources Management, 21,* 333–372.

Stricker, L. J., Rock, D. A., & Burton, N. W. (1993). Sex differences in predictions of college grades from scholastic aptitude test scores. *Journal of Educational Psychology, 4,* 710–718.

Tabachnick, B. G., & Fidell, L. S. (1989). *Using multivariate statistics* (2nd ed.). New York: HarperCollins.

Task Force on Statistical Inference. (2000). Narrow and shallow. *American Psychologist, 55,* 965.

Tate, R. L. (1984). Limitations of centering for interactive models. *Sociological Methods and Research, 13,* 251–271.

Te Nijenhuis, J., & Van der Flier, H. (1999). Bias research in the Netherlands: Review and implications. *European Journal of Psychological Assessment, 15,* 165–175.

Tisak, J. (1994). Determination of the regression coefficients and their associated standard errors in hierarchical regression analysis. *Multivariate Behavioral Research, 29,* 185–201.

Toops, H. A. (1959). A research utopia in industrial psychology. *Personnel Psychology, 12,* 189–225.

Tryon, W. W. (2001). Evaluating statistical difference, equivalence, and indeterminacy using inferential confidence intervals: An integrated alternative method of conducting null hypothesis statistical tests. *Psychological Methods, 6,* 371–386.

Tubbs, M. E. (1993). Commitment as a moderator of the goal-performance relation: A case for clearer construct definition. *Journal of Applied Psychology, 78,* 86–97.

Tukey, J. W. (1977). *Exploratory data analysis.* Reading, MA: Addison-Wesley.

Turban, D. B., Lau, C., Ngo, H., Chow, I. H. S., & Si, S. X. (2001). Organizational attractiveness of firms in the People's Republic of China: A person–organization fit perspective. *Journal of Applied Psychology, 86,* 194–206.

Vargha, A., Rudas, T., Delaney, H. D., & Maxwell, S. E. (1996). Dichotomization, partial correlation, and conditional independence. *Journal of Educational and Behavioral Statistics, 21*, 267–284.

Vasse, R. M., Nijhuis, F. J. N., & Kok, G. (1998). Associations between work stress, alcohol consumption, and sickness absence. *Addiction, 93*, 231–241.

Velicer, W. F. (1972). The moderator variable viewed as heterogeneous regression. *Journal of Applied Psychology, 56*, 266–269.

Villa, J. R., Howell, J. P., Dorfman, P. W., & Daniel, D. L. (2003). Problems with detecting moderators in leadership research using moderated multiple regression. *Leadership Quarterly, 14*, 3–23.

Vohs, K. D., Bardone, A. M., Joiner, T. E., Abramson, L. Y., & Heatherton, T. F. (1999). Perfectionism, perceived weight status, and self-esteem interact to predict bulimic symptoms: A model of bulimic symptom development. *Journal of Abnormal Psychology, 108*, 695–700.

Wainer, H. (1999). One cheer for null hypothesis significance testing. *Psychological Methods, 4*, 212–213.

Wanberg, C. R. (1997). Antecedents and outcomes of coping behaviors among unemployed and reemployed individuals. *Journal of Applied Psychology, 82*, 731–744.

Weinzimmer, L. G., Mone, M. A., & Alwan, L. C. (1994). An examination of perceptions and usage of regression diagnostics in organization studies. *Journal of Management, 20*, 179–192.

West, S. G., Aiken, L. S., & Krull, J. L. (1996). Experimental personality designs: Analyzing categorical by continuous variable interactions. *Journal of Personality, 64*, 1–48.

White, J. K. (1978). Individual differences and the job quality–worker response relationship: Review, integration, and comments. *Academy of Management Review, 3*, 267–280.

White, P. F., & Piette, M. J. (1998). The use of "reverse regression" in employment discrimination analysis. *Journal of Forensic Economics, 11*, 127–138.

Wicherski, M., Pate, W. E., & Kohout, J. (2001). 2000–2001 faculty salaries in graduate departments of psychology. Available online at *http://research. apa.org/facsals0001.html.*

Wilcox, R. R. (1988). A new alternative to the ANOVA F and new results on James's second-order method. *British Journal of Mathematical and Statistical Psychology, 41*, 109–117.

Wilcox, R. R. (1997a). A bootstrap modification of the Alexander–Govern ANOVA method, plus comments on comparing trimmed means. *Educational and Psychological Measurement, 57*, 655–665.

Wilcox, R. R. (1997b). Comparing the slopes of two independent regression lines when there is complete heteroscedasticity. *British Journal of Mathematical and Statistical Psychology, 50*, 309–317.

Wilkinson, L., & the Task Force on Statistical Inference (1999). Statistical methods in psychology journals: Guidelines and explanations. *American Psychologist, 54*, 594–604.

Williams, K. J., & Alliger, G. M. (1994). Role stressors, mood spillover, and perceptions of work-family conflict in employed parents. *Academy of Management Journal, 37*, 837–868.

Winer, B. J. (1974). *Statistical principles in experimental design.* New York: McGraw-Hill.

Winer, B. J., Brown, D. R., & Michels, K. M. (1991). *Statistical principles in experimental design* (3rd ed.). New York: McGraw-Hill.

Wise, S. L., Peters, L. H., & O'Connor, E. J. (1984). Identifying moderator variables using multiple regression: A reply to Darrow & Kahl. *Journal of Management, 10,* 227–236.

Witt, L. A., & Nye, L. G. (1998, April). *Effect sizes in moderated multiple regression: Beyond the increment in R^2.* Paper presented at the meeting of the Society for Industrial and Organizational Psychology, Dallas, TX.

Young, J. W. (1994). Differential prediction of college grades by gender and ethnicity: A replication study. *Educational and Psychological Measurement, 54,* 1022–1029.

Zedeck, S. (1971). Problems with the use of "moderator" variables. *Psychological Bulletin, 76,* 295–310.

Author Index

Subject Index

A priori, 7, 8, 150, 151
A statistic, 57, 60, 61, 62, 170
Abilities, 11, 12, 17, 71, 73, 81
Absenteeism, measure of, 12
Academy of Management Journal, 19, 20*f*, 47, 51, 52, 78, 80, 142
African Americans, 51, 76, 93, 118, 120, 121, 122, 123
Age, 2, 52, 82
  groups, 2, 82
Alcohol, use, 5, 6
Alexander's normalized-*t* approximation, 56, 61
Alpha, 113
ALTMMR, 58, 59, 60, 61, 62, 62*f*, 63, 64, 136, 159
American Psychological Association, 17, 114–115
  journals, 114–115
Analysis of covariance, 19
Analysis of variance
  homogeneity of variance assumption, 43, 54
  one-way, 135
Anxiety, 6
Applicants, 13, 71, 88
Applied psychology, 19, 52, 72, 78, 82, 114
  research, 19, 72, 78, 82
Aptitude, 82. *See also* Scholastic aptitude
  by treatment interactions, 18
  mathematical, 126, 128, 128*f*, 129
ASCII file, 91
Assumption, statistical, 20, 41, 50, 53, 54, 55, 58, 59

Baby Boomers, 82
Backward regression technique, 16
Bartlett's *M* statistic, 54–55, 56, 60, 167
Base salary, 23, 24
Bias, 2, 16, 18, 41, 48, 88, 89, 94, 151
Bootstrapping, 72, 89, 90, 96*t*
Boundary conditions, 4, 155
Brand
  attitude, 2
  loyalty, 2

Causal antecedent, 3
Causal effect, 4
Causal law, 4
Causal relation, 5, 155
Causation, 3
  joint, 3
Centering, 36, 37, 38, 41, 77, 132, 136, 147, 158, 165
Central tendency, 79
Charts, 25, 27*f*
Chi-square distribution, 61
China, 1, 3, 7, 8
Chinese sample, 7
Clinical psychology literature, 18
Coding, 25, 33, 34, 39, 41, 53, 118–124, 135–136
Coding scheme, 12, 13, 15, 116, 117, 122, 125, 135, 136, 154, 158, 164, 165
  contrast, 119, 120*t*, 123–124, 125, 127*f*, 135, 158
  dummy, 24, 33, 34, 37, 38, 39, 41, 118, 119–121, 124, 132, 135, 150, 157, 158
  unweighted effect, 119, 120*t*, 121–122, 135, 158

194